GCSE
Physics

GCSE Physics isn't the easiest subject on Earth, but we've squeezed the important facts, theory and practical skills you'll need into this brilliant all-in-one CGP book!

We've also included plenty of exam-style practice in every topic *and* a full set of practice papers so that you can put your new-found Physics knowledge to the test.

What's more, there are step-by-step answers at the back, so you can easily check your work and find out how to pick up any marks you missed out on!

How to access your free Online Edition

This book includes a free Online Edition to read on your PC, Mac or tablet.
You'll just need to go to **cgpbooks.co.uk/extras** and enter this code:

0186 2671 4074 5504

By the way, this code only works for one person. If somebody else has used this book before you, they might have already claimed the Online Edition.

Complete
Revision & Practice
Everything you need to pass the exams!

Contents

Throughout this book you'll see grade stamps like these:

Grade 4-6   **Grade 6-7**   **Grade 7-9**

These grade stamps help to show how difficult the questions are.

Remember — to get a top grade you need to be able to answer **all** the questions, not just the hardest ones.

In the real exams, some questions test how well you can write (as well as your scientific knowledge).
In this book, we've marked these questions with an asterisk (*).

Topic 4 — Atomic Structure

Topic 5 — Forces

Topic 6 — Waves

Published by CGP

Editors: Sharon Keeley-Holden, Duncan Lindsay, Sophie Scott and Charlotte Whiteley.

Contributors: Paddy Gannon, Gemma Hallam, Jason Howell, Barbara Mascetti.

From original material by Richard Parsons.

With thanks to Emily Garrett, Frances Rooney and Mark A. Edwards for the proofreading.

With thanks to Ana Pungartnik for the copyright research.

Data used to construct stopping distance diagram on page 119 from the Highway Code.
Contains public sector information licensed under the Open Government Licence v3.0.
http://www.nationalarchives.gov.uk/doc/open-government-licence/version/3/

Renewables data on page 198 contain public sector information licensed under the Open Government Licence v3.0.
http://www.nationalarchives.gov.uk/doc/open-government-licence/version/3/

Data used to construct braking distances data diagram on page 218 contains
public sector information licensed under the Open Government Licence v3.0.
http://www.nationalarchives.gov.uk/doc/open-government-licence/version/3/

Printed by Elanders Ltd, Newcastle upon Tyne.
Clipart from Corel®
Illustrations by: Sandy Gardner Artist, email sandy@sandygardner.co.uk

The Scientific Method

This section isn't about how to 'do' science — but it does show you the way most scientists work.

Scientists Come Up With **Hypotheses** — Then **Test** Them

1) Scientists try to explain things. They start by observing something they don't understand.

2) They then come up with a hypothesis — a possible explanation for what they've observed.

3) The next step is to test whether the hypothesis might be right or not. This involves making a prediction based on the hypothesis and testing it by gathering evidence (i.e. data) from investigations. If evidence from experiments backs up a prediction, you're a step closer to figuring out if the hypothesis is true.

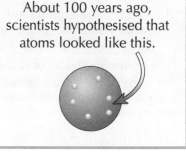

About 100 years ago, scientists hypothesised that atoms looked like this.

Several Scientists Will **Test** a Hypothesis

1) Normally, scientists share their findings in peer-reviewed journals, or at conferences.

2) Peer-review is where other scientists check results and scientific explanations to make sure they're 'scientific' (e.g. that experiments have been done in a sensible way) before they're published. It helps to detect false claims, but it doesn't mean that findings are correct — just that they're not wrong in any obvious way.

3) Once other scientists have found out about a hypothesis, they'll start basing their own predictions on it and carry out their own experiments. They'll also try to reproduce the original experiments to check the results — and if all the experiments in the world back up the hypothesis, then scientists start to think the hypothesis is true.

4) However, if a scientist does an experiment that doesn't fit with the hypothesis (and other scientists can reproduce the results) then the hypothesis may need to be modified or scrapped altogether.

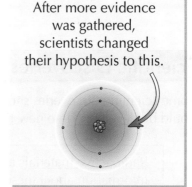

After more evidence was gathered, scientists changed their hypothesis to this.

If **All** the **Evidence** Supports a Hypothesis, It's **Accepted** — For Now

1) Accepted hypotheses are often referred to as theories. Our currently accepted theories are the ones that have survived this 'trial by evidence' — they've been tested many times over the years and survived.

2) However, theories never become totally indisputable fact. If new evidence comes along that can't be explained using the existing theory, then the hypothesising and testing is likely to start all over again.

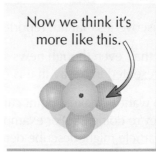

Now we think it's more like this.

Scientific models are constantly being refined...

The scientific method has been developed over time. Aristotle (a Greek philosopher) was the first person to realise that theories need to be based on observations. Muslim scholars then introduced the ideas of creating a hypothesis, testing it, and repeating work to check results.

Models and Communication

Once scientists have made a <u>new discovery</u>, they <u>don't</u> just keep it to themselves. Oh no. Time to learn about how scientific discoveries are <u>communicated</u>, and the <u>models</u> that are used to represent theories.

Theories Can Involve **Different Types** of **Models**

1) A <u>representational model</u> is a <u>simplified description</u> or <u>picture</u> of what's going on in real life. Like all models, it can be used to <u>explain</u> <u>observations</u> and <u>make predictions</u>. E.g. the <u>Bohr model</u> of an atom is a simplified way of showing the arrangement of electrons in an atom (see p.73). It can be used to explain electron excitations in atoms.

Scientists test models by carrying out experiments to check that the predictions made by the model happen as expected.

2) <u>Computational models</u> use computers to make <u>simulations</u> of complex real-life processes, such as climate change. They're used when there are a <u>lot</u> of different <u>variables</u> (factors that change) to consider, and because you can easily <u>change their design</u> to take into account <u>new data</u>.

3) All models have <u>limitations</u> on what they can <u>explain</u> or <u>predict</u>.
E.g. <u>the Big Bang model</u> (a model used to describe the beginning of the Universe) can be used to explain why everything in the Universe is moving away from us. One of its limitations is that it <u>doesn't explain</u> the moments before the Big Bang.

Scientific Discoveries are **Communicated** to the **General Public**

Some scientific discoveries show that people should <u>change their habits</u>, or they might provide ideas that could be <u>developed</u> into new <u>technology</u>. So scientists need to <u>tell the world</u> about their discoveries.

<u>Radioactive materials</u> are used widely in <u>medicine</u> for <u>imaging</u> and <u>treatment</u> (see p.83). Information about these materials needs to be communicated to <u>doctors</u> so they can <u>make use</u> of them, and to <u>patients</u>, so they can make <u>informed decisions</u> about their <u>treatment</u>.

Scientific **Evidence** can be **Presented** in a **Biased Way**

1) Scientific discoveries that are reported in the <u>media</u> (e.g. newspapers or television) <u>aren't</u> peer-reviewed.

2) This means that, even though news stories are often <u>based</u> on data that has been peer-reviewed, the data might be <u>presented</u> in a way that is <u>over-simplified</u> or <u>inaccurate</u>, making it open to <u>misinterpretation</u>.

3) People who want to make a point can sometimes <u>present data</u> in a <u>biased way</u> (sometimes <u>without knowing</u> they're doing it). For example, a scientist might overemphasise a relationship in the data, or a newspaper article might describe details of data <u>supporting</u> an idea without giving any evidence <u>against</u> it.

Companies can present biased data to help sell products...

Sometimes a company may only want you to see half of the story so they present the data in a <u>biased way</u>. For example, a pharmaceutical company may want to encourage you to buy their drugs by telling you about all the <u>positives</u>, but not report the results of any <u>unfavourable studies</u>.

Issues Created by Science

Science has helped us make progress in loads of areas, from advances in medicine to space travel. But science still has its issues. And it can't answer everything, as you're about to find out.

Scientific Developments are Great, but they can Raise Issues

Scientific knowledge is increased by doing experiments. And this knowledge leads to scientific developments, e.g. new technologies or new advice. These developments can create issues though. For example:

Economic issues: Society can't always afford to do things scientists recommend (e.g. investing in alternative energy sources) without cutting back elsewhere.

Social issues: Decisions based on scientific evidence affect people — e.g. should fossil fuels be taxed more highly? Would the effect on people's lifestyles be acceptable?

Personal issues: Some decisions will affect individuals. For example, someone might support alternative energy, but object if a wind farm was built next to their house.

Environmental issues: Human activity often affects the natural environment. For example, building a dam to produce electricity will change the local habitat so some species might be displaced. But it will also reduce our need for fossil fuels, so will help to reduce climate change.

Science Can't Answer Every Question — Especially Ethical Ones

1) We don't understand everything. We're always finding out more, but we'll never know all the answers.

2) In order to answer scientific questions, scientists need data to provide evidence for their hypotheses.

3) Some questions can't be answered yet because the data can't currently be collected, or because there's not enough data to support a theory.

4) Eventually, as we get more evidence, we'll answer some of the questions that currently can't be answered, e.g. what the impact of global warming on sea levels will be. But there will always be the "Should we be doing this at all?"-type questions that experiments can't help us to answer...

Think about new drugs which can be taken to boost your 'brain power'.

- Some people think they're good as they could improve concentration or memory. New drugs could let people think in ways beyond the powers of normal brains.

- Other people say they're bad — they could give some people an unfair advantage in exams. And people might be pressured into taking them so that they could work more effectively, and for longer hours.

There are often issues with new scientific developments...

The trouble is, there's often no clear right answer where these issues are concerned. Different people have different views, depending on their priorities. These issues are full of grey areas.

Risk

Scientific discoveries are often great, but they can prove risky. With dangers all around, you've got to be aware of hazards — this includes how likely they are to cause harm and how serious the effects may be.

Nothing is Completely **Risk-Free**

1) A hazard is something that could potentially cause harm.

2) All hazards have a risk attached to them — this is the chance that the hazard will cause harm.

3) The risks of some things seem pretty obvious, or we've known about them for a while, like the risk of causing acid rain by polluting the atmosphere, or of having a car accident when you're travelling in a car.

4) New technology arising from scientific advances can bring new risks, e.g. scientists are unsure whether nanoparticles that are being used in cosmetics and suncream might be harming the cells in our bodies. These risks need to be considered alongside the benefits of the technology, e.g. improved sun protection.

5) You can estimate the size of a risk based on how many times something happens in a big sample (e.g. 100 000 people) over a given period (e.g. a year). For example, you could assess the risk of a driver crashing by recording how many people in a group of 100 000 drivers crashed their cars over a year.

6) To make decisions about activities that involve a risk, we need to take into account the chance of the hazard causing harm, and how serious the consequences would be if it did. If an activity involves a risk that's very likely to cause harm, with serious consequences if it does, that activity is considered high risk.

People Make Their **Own Decisions** About Risk

1) Not all risks have the same consequences, e.g. if you chop veg with a sharp knife you risk cutting your finger, but if you go scuba-diving you risk death. You're much more likely to cut your finger during half an hour of chopping than to die during half an hour of scuba-diving. But most people are happier to accept a higher probability of an accident if the consequences are short-lived and fairly minor.

2) People tend to be more willing to accept a risk if they choose to do something (e.g. go scuba diving), compared to having the risk imposed on them (e.g. having a nuclear power station built next door).

3) People's perception of risk (how risky they think something is) isn't always accurate. They tend to view familiar activities as low-risk and unfamiliar activities as high-risk — even if that's not the case. For example, cycling on roads is often high-risk, but many people are happy to do it because it's a familiar activity. Air travel is actually pretty safe, but a lot of people perceive it as high-risk.

4) People may underestimate the risk of things with long-term or invisible effects, e.g. using tanning beds.

The pros and cons of new technology must be weighed up...

The world's a dangerous place and it's impossible to rule out the chance of an accident altogether. But if you can recognise hazards and take steps to reduce the risks, you're more likely to stay safe.

Designing Investigations

Dig out your lab coat and dust off your badly-scratched safety goggles... it's <u>investigation time</u>.

Evidence Can Support or Disprove a Hypothesis

1) Scientists <u>observe</u> things and come up with <u>hypotheses</u> to test them (see p.1).
You need to be able to do the same. For example:

> <u>Observation</u>: People with big feet have spots. <u>Hypothesis</u>: Having big feet causes spots.

2) To <u>determine</u> whether or not a hypothesis is <u>right</u>, you need to do an
<u>investigation</u> to gather evidence. To do this, you need to use your hypothesis
to make a <u>prediction</u> — something you think <u>will happen</u> that you can test.
E.g. people who have bigger feet will have more spots.

Investigations include experiments and studies.

3) Investigations are used to see if there are <u>patterns</u> or <u>relationships</u>
between <u>two variables</u>, e.g. to see if there's a pattern or relationship
between the variables 'number of spots' and 'size of feet'.

Evidence Needs to be Repeatable, Reproducible and Valid

1) <u>Repeatable</u> means that if the <u>same person</u> does an experiment again
using the <u>same methods</u> and equipment, they'll get <u>similar results</u>.

2) <u>Reproducible</u> means that if <u>someone else</u> does the experiment, or a <u>different</u>
method or piece of equipment is used, the results will still be <u>similar</u>.

3) If data is <u>repeatable</u> and <u>reproducible</u>, it's <u>reliable</u> and
scientists are more likely to <u>have confidence</u> in it.

4) <u>Valid results</u> are both repeatable and reproducible AND they <u>answer the original question</u>.
They come from experiments that were designed to be a <u>FAIR TEST</u>...

Make an Investigation a Fair Test By Controlling the Variables

1) In a lab experiment you usually <u>change one variable</u> and <u>measure</u> how it affects <u>another variable</u>.

2) To make it a fair test, <u>everything else</u> that could affect the results should <u>stay the same</u>
— otherwise you can't tell if the thing you're changing is causing the results or not.

3) The variable you <u>CHANGE</u> is called the <u>INDEPENDENT</u> variable.

4) The variable you <u>MEASURE</u> when you change the independent variable is the <u>DEPENDENT</u> variable.

5) The variables that you <u>KEEP THE SAME</u> are called <u>CONTROL</u> variables.

> You could find how <u>current</u> through a circuit component affects the <u>potential difference</u>
> across the component by measuring the <u>potential difference</u> at different currents.
> The <u>independent variable</u> is the <u>current</u>. The <u>dependent variable</u> is the <u>potential difference</u>.
> <u>Control variables</u> include the <u>temperature</u> of the component, the <u>pd</u> of the power supply, etc.

6) Because you can't always control all the variables, you often need to use a <u>control experiment</u>. This is
an experiment that's kept under the <u>same conditions</u> as the rest of the investigation, but <u>doesn't</u> have
anything <u>done</u> to it. This is so that you can see what happens when you don't change anything at all.

Designing Investigations

The **Bigger** the **Sample Size** the **Better**

1) Data based on small samples isn't as good as data based on large samples. A sample should represent the whole population (i.e. it should share as many of the characteristics in the population as possible) — a small sample can't do that as well. It's also harder to spot anomalies if your sample size is too small.

2) The bigger the sample size the better, but scientists have to be realistic when choosing how big. For example, if you were studying the effects of living near a nuclear power plant, it'd be great to study everyone who lived near a nuclear power plant (a huge sample), but it'd take ages and cost a bomb. It's more realistic to study a thousand people, with a range of ages and races and across both genders.

Your **Equipment** has to be **Right for the Job**

1) The measuring equipment you use has to be sensitive enough to measure the changes you're looking for. For example, if you need to measure changes of 1 cm^3 you need to use a measuring cylinder that can measure in 1 cm^3 steps — it'd be no good trying with one that only measures 10 cm^3 steps.

2) The smallest change a measuring instrument can detect is called its resolution. E.g. some mass balances have a resolution of 1 g, some have a resolution of 0.1 g, and some are even more sensitive.

3) Also, equipment needs to be calibrated by measuring a known value. If there's a difference between the measured and known value, you can use this to adjust your measurements to compensate for the inaccuracy of the equipment.

Data Should be **Repeatable, Reproducible, Accurate** and **Precise**

1) To check repeatability you need to repeat the readings and check that the results are similar. You need to repeat each reading at least three times.

2) To make sure your results are reproducible you can cross check them by taking a second set of readings with another instrument (or a different observer).

3) Your data also needs to be accurate. Really accurate results are those that are really close to the true answer. The accuracy of your results usually depends on your method — you need to make sure you're measuring the right thing and that you don't miss anything that should be included in the measurements. E.g. estimating the volume of an irregularly shaped solid by measuring the sides isn't very accurate because this will not take into account any gaps in the object. It's more accurate to measure the volume using a eureka can (see p.181).

Repeat	Data set 1	Data set 2
1	12	11
2	14	17
3	13	14
Mean	13	14

Data set 1 is more precise than data set 2.

4) Your data also needs to be precise. Precise results are ones where the data is all really close to the mean (average) of your repeated results (i.e. not spread out).

Designing Investigations

You Need to Look out for **Errors** and **Anomalous Results**

1) The results of your experiment will always <u>vary a bit</u> because of <u>random errors</u> — unpredictable differences caused by things like <u>human errors</u> in <u>measuring</u>. The errors when you make a reading from a ruler are random. You have to estimate or round the distance when it's between two marks — so sometimes your figure will be a bit above the real one, and sometimes it will be a bit below.

2) You can <u>reduce</u> the effect of random errors by taking <u>repeat readings</u> and finding the <u>mean</u>. This will make your results <u>more precise</u>.

If there's no systematic error, then doing repeats and calculating a mean could make your results more accurate.

3) If a measurement is wrong by the <u>same amount every time</u>, it's called a <u>systematic error</u>. For example, if you measured from the very end of your ruler instead of from the 0 cm mark every time, all your measurements would be a bit small. Repeating the experiment in the exact same way and calculating a mean <u>won't</u> correct a systematic error.

4) Just to make things more complicated, if a systematic error is caused by using <u>equipment</u> that <u>isn't zeroed properly</u>, it's called a <u>zero error</u>. For example, if a mass balance always reads 1 gram before you put anything on it, all your measurements will be 1 gram too heavy.

5) You can <u>compensate</u> for some systematic errors if you know about them, e.g. if a mass balance always reads 1 gram before you put anything on it, you can subtract 1 gram from all your results.

6) Sometimes you get a result that <u>doesn't fit in</u> with the rest at all. This is called an <u>anomalous result</u>. You should investigate it and try to <u>work out what happened</u>. If you can work out what happened (e.g. you measured something wrong) you can <u>ignore</u> it when processing your results.

Investigations Can be **Hazardous**

1) <u>Hazards</u> from science experiments might include:

> - <u>Lasers</u>, e.g. if a laser is directed into the eye, this can cause blindness.
> - <u>Gamma radiation</u>, e.g. gamma-emitting radioactive sources can cause cancer.
> - <u>Fire</u>, e.g. an unattended Bunsen burner is a fire hazard.
> - <u>Electricity</u>, e.g. faulty electrical equipment could give you a shock.

You can find out about potential hazards by looking in textbooks, doing some internet research, or asking your teacher.

2) Part of planning an investigation is making sure that it's <u>safe</u>.

3) You should always make sure that you <u>identify</u> all the hazards that you might encounter. Then you should think of ways of <u>reducing the risks</u> from the hazards you've identified. For example:

> - If you're working with <u>springs</u>, always wear safety goggles.
>   This will reduce the risk of the spring hitting your eye if the spring snaps.
> - If you're using a <u>Bunsen burner</u>, stand it on a heat proof mat to reduce the risk of starting a fire.

Designing an investigation is an involved process...

<u>Collecting data</u> is what investigations are all about. Designing a good investigation is really important to make sure that any data collected is <u>accurate</u>, <u>precise</u>, <u>repeatable</u> and <u>reproducible</u>.

Processing Data

Processing your data means doing some <u>calculations</u> with it to make it <u>more useful</u>.

Data Needs to be Organised

1) Tables are really useful for <u>organising data</u>.

2) When you draw a table <u>use a ruler</u> and make sure <u>each column</u> has a <u>heading</u> (including the <u>units</u>).

There are Different Ways to Process Your Data

1) When you've done repeats of an experiment you should always calculate the <u>mean</u> (average). To do this <u>add together</u> all the data values and <u>divide</u> by the total number of values in the sample.

2) You can also find the <u>mode</u> of your results — this is the <u>value</u> that <u>occurs</u> the <u>most</u> in your set of results.

3) The <u>median</u> can be found by writing your results in numerical <u>order</u> — the median is the <u>middle number</u>.

Ignore anomalous results when calculating the mean, mode and median.

EXAMPLE: **The results of an experiment show the extension of two springs when a force is applied to both of them. Calculate the mean, mode and median of the extension for both springs.**

Spring	Repeat (cm)					Mean (cm)	Mode (cm)	Median (cm)
	1	2	3	4	5			
A	18	26	22	26	28	(18 + 26 + 22 + 26 + 28) ÷ 5 = 24	26	26
B	11	14	20	15	20	(11 + 14 + 20 + 15 + 20) ÷ 5 = 16	20	15

Round to the Lowest Number of Significant Figures

The <u>first significant figure</u> of a number is the first digit that's <u>not zero</u>. The second and third significant figures come <u>straight after</u> (even if they're zeros). You should be aware of significant figures in calculations.

1) In <u>any</u> calculation where you need to round, you should round the answer to the <u>lowest number of significant figures</u> (s.f.) given.

2) Remember to write down <u>how many</u> significant figures you've rounded to after your answer.

3) If your calculation has multiple steps, <u>only</u> round the <u>final</u> answer, or it won't be as accurate.

EXAMPLE: **The mass of a solid is 0.24 g and its volume is 0.715 cm³. Calculate the density of the solid.**

Density = 0.24 g ÷ 0.715 cm³ = 0.33566... = 0.34 g/cm³ (2 s.f.)

2 s.f.   3 s.f.   Final answer should be rounded to 2 s.f.

Don't forget your calculator...

In the exam you could be given some <u>data</u> and be expected to <u>process it</u> in some way. Make sure you keep an eye on <u>significant figures</u> in your answers and <u>always write down your working</u>.

Presenting Data

Once you've processed your data, e.g. by calculating the mean, you can present your results in a nice <u>chart</u> or <u>graph</u>. This will help you to <u>spot any patterns</u> in your data.

If Your Data Comes in **Categories**, Present It in a **Bar Chart**

1) If the independent variable is <u>categoric</u> (comes in distinct categories, e.g. solid, liquid, gas) you should use a <u>bar chart</u> to display the data.

2) You also use them if the independent variable is <u>discrete</u> (the data can be counted in chunks, where there's no in-between value, e.g. number of protons is discrete because you can't have half a proton).

3) There are some <u>golden rules</u> you need to follow for <u>drawing</u> bar charts:

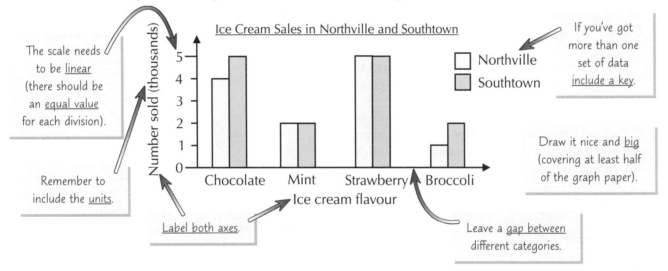

The scale needs to be <u>linear</u> (there should be an <u>equal value</u> for each division).

Remember to include the <u>units</u>.

Label both axes.

If you've got more than one set of data <u>include a key</u>.

Draw it nice and <u>big</u> (covering at least half of the graph paper).

Leave a <u>gap between</u> different categories.

If Your Data is **Continuous**, Plot a **Graph**

1) If both variables are <u>continuous</u> (numerical data that can have any value within a range, e.g. length, volume, temperature) you should use a <u>graph</u> to display the data.

2) Here are the <u>rules</u> for plotting points on a graph:

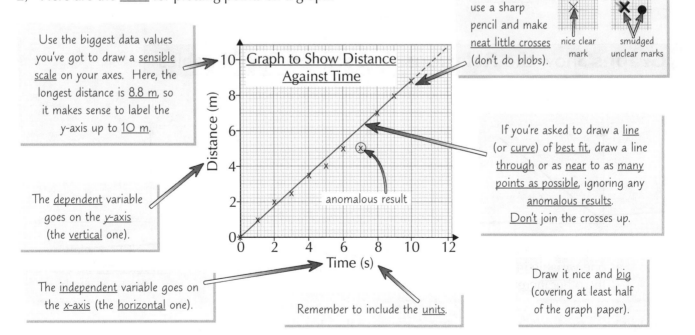

Use the biggest data values you've got to draw a <u>sensible scale</u> on your axes. Here, the longest distance is <u>8.8 m</u>, so it makes sense to label the y-axis up to <u>10 m</u>.

The <u>dependent</u> variable goes on the <u>y-axis</u> (the <u>vertical</u> one).

The <u>independent</u> variable goes on the <u>x-axis</u> (the horizontal one).

To plot points, use a sharp pencil and make <u>neat little crosses</u> (don't do blobs). nice clear mark / smudged unclear marks

If you're asked to draw a <u>line</u> (or <u>curve</u>) of <u>best fit</u>, draw a line <u>through</u> or as <u>near</u> to as <u>many points as possible</u>, ignoring any <u>anomalous results</u>. <u>Don't</u> join the crosses up.

Remember to include the <u>units</u>.

Draw it nice and <u>big</u> (covering at least half of the graph paper).

More on Graphs

Graphs aren't just fun to plot, they're also really useful for showing <u>trends</u> in your data.

Graphs Can Give You a Lot of Information About Your Data

1) The <u>gradient</u> (slope) of a graph tells you how quickly the <u>dependent variable</u> changes if you change the <u>independent variable</u>.

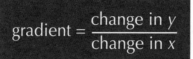

$$\text{gradient} = \frac{\text{change in } y}{\text{change in } x}$$

You can use this method to calculate other rates from a graph, so long as you always make x time.

This <u>graph</u> shows the <u>distance travelled</u> by a vehicle against <u>time</u>. The graph is <u>linear</u> (it's a straight line graph), so you can simply calculate the <u>gradient</u> of the line to find out the <u>speed</u> of the vehicle.

1) To calculate the gradient, pick <u>two points</u> on the line that are easy to read and a <u>good distance</u> apart.

2) <u>Draw a line down</u> from one of the points and a <u>line across</u> from the other to make a <u>triangle</u>. The line drawn down the side of the triangle is the <u>change in y</u> and the line across the bottom is the <u>change in x</u>.

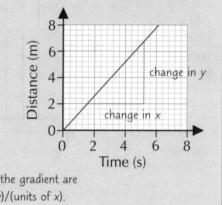

Change in y = 6.8 − 2.0 = 4.8 m   Change in x = 5.2 − 1.6 = 3.6 s

Rate = gradient = $\frac{\text{change in } y}{\text{change in } x}$ = $\frac{4.8 \text{ m}}{3.6 \text{ s}}$ = <u>1.3 m/s</u>   The units of the gradient are (units of y)/(units of x).

2) To find the <u>gradient of a curve</u> at a <u>certain point</u>, draw a <u>tangent</u> to the curve at that point. This is a <u>straight line</u> that <u>touches</u> the curve at that <u>point</u>, but doesn't <u>cross</u> it. Then just find the <u>gradient of the tangent</u> in the same way as above.

3) The <u>intercept</u> of a graph is where the line of best fit crosses one of the <u>axes</u>. The <u>x-intercept</u> is where the line of best fit crosses the x-axis and the <u>y-intercept</u> is where it crosses the <u>y-axis</u>.

Graphs Show the Relationship Between Two Variables

1) You can get <u>three</u> types of <u>correlation</u> (relationship) between variables:

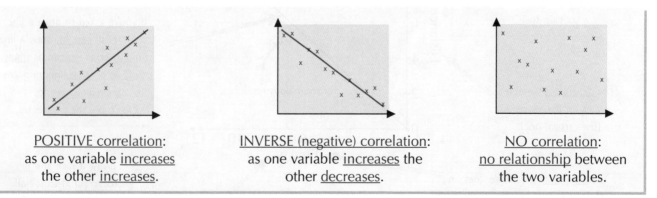

<u>POSITIVE correlation:</u> as one variable <u>increases</u> the other <u>increases</u>.

<u>INVERSE (negative) correlation:</u> as one variable <u>increases</u> the other <u>decreases</u>.

<u>NO correlation:</u> <u>no relationship</u> between the two variables.

2) Just because there's correlation, it doesn't mean the change in one variable is <u>causing</u> the change in the other — there might be <u>other factors</u> involved (see page 14).

Units

Graphs and maths skills are all very well, but the numbers don't mean much if you don't get the <u>units</u> right.

S.I. Units Are Used All Round the World

1) It wouldn't be all that useful if I defined volume in terms of <u>bath tubs</u>, you defined it in terms of <u>egg-cups</u> and my pal Fred defined it in terms of <u>balloons</u> — we'd never be able to compare our data.

2) To stop this happening, scientists have come up with a set of <u>standard units</u>, called S.I. units, that all scientists use to measure their data. Here are some S.I. units you'll see in physics:

Quantity	S.I. Base Unit
mass	kilogram, kg
length	metre, m
time	second, s
temperature	kelvin, K

Always Check The Values Used in Equations Have the Right Units

1) Formulas and equations show <u>relationships</u> between <u>variables</u>.

2) To <u>rearrange</u> an equation, make sure that whatever you do to <u>one side</u> of the equation you also do to the <u>other side</u>.

For example, you can find the <u>speed</u> of a wave using the equation: ⟹ wave speed = frequency × wavelength

You can <u>rearrange</u> this equation to find the <u>frequency</u> by <u>dividing</u> <u>each side</u> by wavelength to give: ⟹ frequency = wave speed ÷ wavelength

3) To use a formula, you need to know the values of <u>all but one</u> of the variables. <u>Substitute</u> the values you do know into the formula, and do the calculation to work out the final variable.

4) Always make sure the values you put into an equation or formula have the <u>right units</u>. For example, you might have done an experiment to find the speed of a trolley. The distance the trolley travels will probably have been measured in cm, but the equation to find speed uses distance in m. So you'll have to <u>convert</u> your distance from cm to m before you put it into the equation.

5) To make sure your units are <u>correct</u>, it can help to write down the <u>units</u> on each line of your <u>calculation</u>.

S.I. units help scientists to compare data...

You can only really <u>compare</u> things if they're in the <u>same units</u>. For example, if you measured the speed of one car in m/s, and one in km/h, it would be hard to know which car was going faster.

Converting Units

You can <u>convert units</u> using <u>scaling prefixes</u>. This can save you from having to write a lot of 0's...

Scaling Prefixes Can Be Used for Large and Small Quantities

1) Quantities come in a huge <u>range</u> of sizes. For example, the volume of a swimming pool might be around 2 000 000 000 cm³, while the volume of a cup is around 250 cm³.

2) To make the size of numbers more <u>manageable</u>, larger or smaller units are used. These are the <u>S.I. base units</u> (e.g. metres) with a <u>prefix</u> in front:

Prefix	tera (T)	giga (G)	mega (M)	kilo (k)	deci (d)	centi (c)	milli (m)	micro (µ)	nano (n)
Multiple of Unit	10^{12}	10^{9}	1 000 000 (10^{6})	1000	0.1	0.01	0.001	0.000001 (10^{-6})	10^{-9}

3) These <u>prefixes</u> tell you <u>how much bigger</u> or <u>smaller</u> a unit is than the base unit. So one <u>kilo</u>metre is <u>one thousand</u> metres.

4) To <u>swap</u> from one unit to another, all you need to know is what number you have to divide or multiply by to get from the original unit to the new unit — this is called the <u>conversion factor</u>.

The conversion factor is the number of times the smaller unit goes into the larger unit.

- To go from a <u>bigger unit</u> (like m) to a <u>smaller unit</u> (like cm), you <u>multiply</u> by the conversion factor.
- To go from a <u>smaller unit</u> (like g) to a <u>bigger unit</u> (like kg), you <u>divide</u> by the conversion factor.

5) Here are some conversions that'll be useful for GCSE physics:

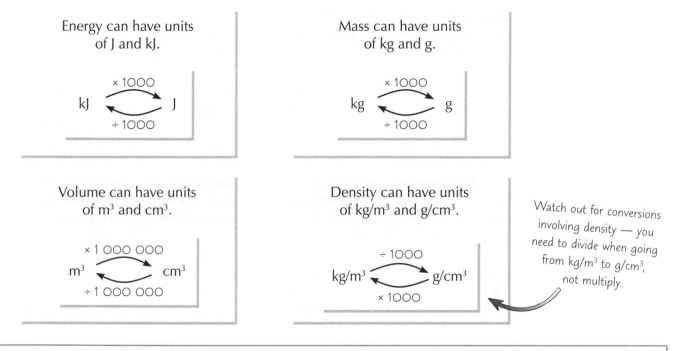

Energy can have units of J and kJ.

Mass can have units of kg and g.

Volume can have units of m³ and cm³.

Density can have units of kg/m³ and g/cm³.

Watch out for conversions involving density — you need to divide when going from kg/m³ to g/cm³, not multiply.

To convert from bigger units to smaller units...

...multiply by the conversion factor. And to convert from <u>smaller units</u> to <u>bigger units</u>, <u>divide</u> by the <u>conversion factor</u>. Don't go getting this rule muddled up and the wrong way round...

Drawing Conclusions

Once you've designed your experiment, carried it out, processed and presented your data, it's finally time to sit down and work out exactly what your data tells you. Time for some fun with conclusions...

You Can **Only Conclude** What the Data Shows and **No More**

1) Drawing conclusions might seem pretty straightforward — you just <u>look at your data</u> and <u>say what pattern or relationship you see</u> between the dependent and independent variables.

The table on the right shows the potential difference across a light bulb for three <u>different</u> currents through the bulb:

Current (A)	Potential difference (V)
6	4
9	10
12	13

CONCLUSION:
A <u>higher current</u> through the bulb gives a higher <u>potential difference</u> across the bulb.

2) But you've got to be really careful that your conclusion <u>matches the data</u> you've got and <u>doesn't go any further</u>.

> You <u>can't</u> conclude that the potential difference across <u>any circuit component</u> will be higher for a larger current — the results might be completely different.

3) You also need to be able to <u>use your results</u> to <u>justify your conclusion</u> (i.e. back up your conclusion with some specific data).

> The potential difference across the bulb was <u>9 V higher</u> with a current of 12 A compared to a current of 6 A.

4) When writing a conclusion you need to <u>refer back</u> to the original hypothesis and say whether the data <u>supports it</u> or not:

> The hypothesis for this experiment might have been that a higher current through the bulb would <u>increase</u> the potential difference across the bulb. If so, the data <u>supports</u> the hypothesis.

You should be able to justify your conclusion with your data...

You should always be able to explain how your data <u>supports</u> your <u>conclusion</u>. It's easy to go too far with conclusions and start making <u>bold claims</u> that your data simply can't back up. When you're drawing conclusions, it's also important that you refer back to your <u>initial hypothesis</u>, the one you made right back at the start of the investigation, to see whether your data supports it or not.

Correlation and Cause

Don't get carried away when you're <u>drawing conclusions</u> — <u>correlation</u> doesn't always mean <u>cause</u>. There could be a few reasons why two variables appear to be linked, as you're about to find out.

Correlation DOES NOT Mean Cause

If two things are correlated (i.e. there's a relationship between them) it <u>doesn't</u> necessarily mean a change in one variable is <u>causing</u> the change in the other — this is <u>REALLY IMPORTANT</u> — <u>DON'T FORGET IT</u>.

There are Three Possible Reasons for a Correlation

1) <u>CHANCE</u>: It might seem strange, but two things can show a correlation purely due to <u>chance</u>.

> For example, one study might find a correlation between people's hair colour and how good they are at frisbee. But other scientists <u>don't</u> get a correlation when they investigate it — the results of the first study are just a <u>fluke</u>.

2) <u>LINKED BY A 3RD VARIABLE</u>: A lot of the time it may <u>look</u> as if a change in one variable is causing a change in the other, but it <u>isn't</u> — a <u>third variable links</u> the two things.

> For example, there's a correlation between <u>water temperature</u> and <u>shark attacks</u>. This isn't because warmer water makes sharks crazy. Instead, they're linked by a third variable — the <u>number of people swimming</u> (more people swim when the water's hotter, and with more people in the water you get more shark attacks).

3) <u>CAUSE</u>: Sometimes a change in one variable does <u>cause</u> a change in the other. You can only conclude that a correlation is due to cause when you've <u>controlled all the variables</u> that could, just could, be affecting the result.

> For example, there's a correlation between <u>smoking</u> and <u>lung cancer</u>. This is because chemicals in tobacco smoke cause lung cancer. This conclusion was only made once <u>other variables</u> (such as age and exposure to other things that cause cancer) had been <u>controlled</u>.

Two variables could appear to be linked by chance...

<u>Correlation</u> doesn't necessarily mean <u>cause</u> — two variables might appear to be linked but it could just be down to <u>chance</u>, or they could be linked by a <u>third variable</u>. When you draw conclusions, make sure you're not jumping to conclusions about cause, and check that you properly <u>consider</u> all the reasons why two variables might appear to be linked.

Uncertainty

Uncertainty is how sure you can really be about your data. There's a little bit of maths to do, and also a formula to learn. But don't worry too much — it's no more than a simple bit of subtraction and division.

Uncertainty is the Amount of Error Your Measurements Might Have

1) When you repeat a measurement, you often get a slightly different figure each time you do it due to random error. This means that each result has some uncertainty to it.

2) The measurements you make will also have some uncertainty in them due to limits in the resolution of the equipment you use (see page 6).

3) This all means that the mean of a set of results will also have some uncertainty to it. You can calculate the uncertainty of a mean result using the equation:

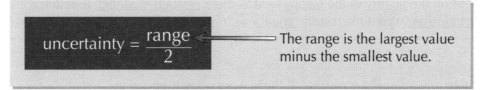

$$\text{uncertainty} = \frac{\text{range}}{2}$$

The range is the largest value minus the smallest value.

4) The larger the range, the less precise your results are and the more uncertainty there will be in your results. Uncertainties are shown using the '±' symbol.

EXAMPLE:

The table below shows the results of an experiment to determine the speed of the trolley as it moves along a horizontal surface. Calculate the uncertainty of the mean.

Repeat	1	2	3	4
Speed (m/s)	2.01	1.98	2.00	2.01

1) First work out the range:

Range = 2.01 − 1.98 = 0.030 m/s

2) Then find the mean:

Mean = (2.01 + 1.98 + 2.00 + 2.01) ÷ 4

= 8.00 ÷ 4 = 2.00

3) Use the range to find the uncertainty:

Uncertainty = range ÷ 2 = 0.030 ÷ 2 = 0.015 m/s

So the uncertainty of the mean = 2.00 ± 0.015 m/s

5) Measuring a greater amount of something helps to reduce uncertainty. For example, in an experiment investigating speed, measuring the distance travelled over a longer period compared to a shorter period will reduce the uncertainty in your results.

The smaller the uncertainty, the more precise your results...

Remember that equation for uncertainty. You never know when you might need it — you could be expected to use it in the exams. You need to make sure all the data is in the same units though. For example, if you had some measurements in metres, and some in centimetres, you'd need to convert them all into either metres or centimetres before you set about calculating uncertainty.

Evaluations

Hurrah! The end of another investigation. Well, now you have to work out all the things you did <u>wrong</u>. That's what <u>evaluations</u> are all about I'm afraid. Best get cracking with this page...

Evaluations — Describe **How** Experiments Could be **Improved**

An evaluation is a <u>critical analysis</u> of the whole investigation.

1) You should comment on the <u>method</u> — was it <u>valid</u>? Did you control all the other variables to make it a <u>fair test</u>?

2) Comment on the <u>quality</u> of the <u>results</u> — was there <u>enough evidence</u> to reach a valid <u>conclusion</u>? Were the results <u>repeatable</u>, <u>reproducible</u>, <u>accurate</u> and <u>precise</u>?

3) Were there any <u>anomalous</u> results? If there were <u>none</u> then <u>say so</u>. If there were any, try to <u>explain</u> them — were they caused by <u>errors</u> in measurement? Were there any other <u>variables</u> that could have <u>affected</u> the results? You should comment on the level of <u>uncertainty</u> in your results too.

4) All this analysis will allow you to say how <u>confident</u> you are that your conclusion is <u>right</u>.

5) Then you can suggest any <u>changes</u> to the <u>method</u> that would <u>improve</u> the quality of the results, so that you could have <u>more confidence</u> in your conclusion. For example, you might suggest <u>changing</u> the way you controlled a variable, or <u>increasing</u> the number of <u>measurements</u> you took. Taking more measurements at <u>narrower intervals</u> could give you a <u>more accurate result</u>. For example:

> <u>Springs</u> have a <u>limit of proportionality</u> (a maximum force before force and extension are no longer proportional). Say you use several <u>identical</u> springs to do an experiment to find the limit of proportionality of the springs. If you apply forces of 1 N, 2 N, 3 N, 4 N and 5 N, and from the results see that it is somewhere <u>between 4 N and 5 N</u>, you could <u>repeat</u> the experiment with one of the other springs, taking <u>more measurements</u> <u>between 4 N and 5 N</u> to get a <u>more accurate</u> value for the limit of proportionality.

6) You could also make more <u>predictions</u> based on your conclusion, then <u>further experiments</u> could be carried out to test them.

When suggesting improvements to the investigation, always make sure that you say why you think this would make the results better.

Always look for ways to improve your investigations...

So there you have it — <u>Working Scientifically</u>. Make sure you know this stuff like the back of your hand. It's not just in the lab, when you're carrying out your groundbreaking <u>investigations</u>, that you'll need to know how to work scientifically. You can be asked about it in the <u>exams</u> as well. So swot up...

Energy Stores

Energy is <u>never used up</u>. Instead it's just <u>transferred</u> between different <u>energy stores</u> and different objects...

Energy is **Transferred** Between **Stores**

When energy is <u>transferred</u> to an object, the energy is <u>stored</u> in one of the object's <u>energy stores</u>. The <u>energy stores</u> you need to know are:

You may also see thermal energy stores called internal energy stores.

1) <u>Thermal</u> energy stores.

2) <u>Kinetic</u> energy stores.

3) <u>Gravitational potential</u> energy stores.

4) <u>Elastic potential</u> energy stores.

5) <u>Chemical</u> energy stores.

6) <u>Magnetic</u> energy stores.

7) <u>Electrostatic</u> energy stores.

8) <u>Nuclear</u> energy stores.

Energy is transferred <u>mechanically</u> (by a <u>force doing work</u>), <u>electrically</u> (work done by a <u>moving charges</u>), by <u>heating</u> (see below) or by <u>radiation</u> (e.g. <u>light</u>, p.136, or <u>sound</u>, p.152).

There's more on doing work on the next page.

When a **System Changes, Energy** is **Transferred**

1) A <u>system</u> is just a fancy word for a <u>single</u> object (e.g. the air in a piston) or a <u>group</u> of <u>objects</u> (e.g. two colliding vehicles) that you're interested in.

2) When a system <u>changes</u>, <u>energy is transferred</u>. It can be transferred <u>into</u> or <u>away from</u> the system, between <u>different objects</u> in the system or between <u>different types</u> of energy stores (e.g. from the kinetic energy store of an object to its thermal energy store).

3) <u>Closed systems</u> are systems where neither <u>matter nor energy can enter or leave</u>. The <u>net change</u> in the <u>total energy</u> of a <u>closed system</u> is <u>always zero</u>.

Energy can be **Transferred** by **Heating**

1) Take the example of <u>boiling water</u> in a <u>kettle</u> — you can think of the <u>water</u> as <u>the system</u>. Energy is <u>transferred to</u> the water (from the kettle's heating element) <u>by heating</u>, into the water's <u>thermal</u> energy store (causing the <u>temperature</u> of the water to <u>rise</u>).

2) You could also think of the <u>kettle's</u> heating element and the <u>water</u> together as a <u>two-object system</u>. Energy is transferred <u>electrically</u> to the <u>thermal</u> energy store of the kettle's heating element, which transfers energy <u>by heating</u> to the water's <u>thermal</u> energy store.

No matter what store it's in, it's all energy...

In the exam, make sure you refer to <u>energy</u> in terms of the <u>store</u> it's in. For example, if you're describing energy in a <u>hot object</u>, say it '<u>has energy in its thermal energy store</u>'.

Work Done

On the previous page, you saw how energy can be <u>transferred</u> between <u>energy stores</u> by <u>heating</u>. Well, that was just the start... This page is all about how energy is transferred when <u>work is done</u>.

Energy can be Transferred by Doing Work

1) <u>Work done</u> is just another way of saying <u>energy transferred</u> — they're the <u>same thing</u>.

2) <u>Work</u> can be done when <u>current flows</u> (work is done <u>against resistance</u> in a <u>circuit</u>, see page 41) or by a <u>force</u> moving an object (there's more on this on page 90). Here are a few examples:

The <u>initial force</u> exerted by a person to <u>throw</u> a ball <u>upwards</u> does <u>work</u>. It causes an energy transfer <u>from</u> the <u>chemical energy store</u> of the person's arm to the <u>kinetic</u> energy store of the ball and arm.

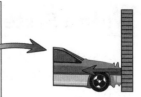

an upwards force is exerted on the ball

The <u>friction</u> between a car's <u>brakes</u> and its <u>wheels</u> does work as the car <u>slows down</u>. It causes an energy transfer from the <u>wheels' kinetic energy</u> stores to the <u>thermal</u> energy store of the <u>surroundings</u>.

frictional forces cause a transfer of energy

In a collision between a car and a <u>stationary object</u>, the <u>normal contact force</u> between the car and the object <u>does work</u>. It causes energy to be transferred from the car's <u>kinetic</u> energy store to <u>other energy stores</u>, e.g. the <u>elastic potential</u> and <u>thermal</u> energy stores of the object and the car body. Some energy might also be <u>transferred away</u> by <u>sound waves</u> (see page 152).

normal contact force causes a transfer of energy to the car

Falling Objects Also Transfer Energy

1) When something, e.g. a ball, is <u>dropped</u> from a height, it's accelerated by <u>gravity</u>. The <u>gravitational force</u> does <u>work</u>.

2) As it <u>falls</u>, energy from the object's <u>gravitational potential energy</u> (g.p.e) <u>store</u> is transferred to its <u>kinetic energy store</u>.

3) For a falling object when there's <u>no air resistance</u>:

gravitational force

Energy <u>lost</u> from the <u>g.p.e. store</u> = Energy <u>gained</u> in the <u>kinetic energy store</u>

4) In real life, <u>air resistance</u> (p.108) acts against all falling objects — it causes some energy to be transferred to <u>other energy stores</u>, e.g. the <u>thermal</u> energy stores of the <u>object</u> and <u>surroundings</u>.

Energy is transferred between the different stores of objects...

Energy stores pop up <u>everywhere</u> in physics. You need to be able to describe <u>how energy is transferred</u>, and <u>which stores</u> it gets transferred between, for <u>any scenario</u>. So, it's time to make sure you know all the <u>energy stores</u> and <u>transfer methods</u> like the back of your hand.

Kinetic and Potential Energy Stores

Now you've got your head around <u>energy stores</u>, it's time to see how you can calculate the amount of energy in <u>three</u> of the most common ones — <u>kinetic</u>, <u>gravitational potential</u> and <u>elastic potential</u> energy stores.

Movement Means Energy in an Object's Kinetic Energy Store

1) Anything that is <u>moving</u> has energy in its <u>kinetic energy store</u>. Energy is transferred <u>to</u> this store when an object <u>speeds up</u> and is transferred <u>away</u> from this store when an object <u>slows down</u>.

2) The energy in the <u>kinetic energy store</u> depends on the object's <u>mass</u> and <u>speed</u>. The <u>greater its mass</u> and the <u>faster</u> it's going, the <u>more energy</u> there will be in its kinetic energy store.

3) There's a <u>slightly tricky</u> formula for it, so you have to concentrate <u>a little bit harder</u> for this one.

Kinetic energy (J) — $E_k = \frac{1}{2}mv^2$ — (Speed)^2 (m/s)^2

Mass (kg)

$\frac{1}{2}mv^2$ means $\frac{1}{2} \times m \times v^2$.

A car of mass 2500 kg is travelling at 20 m/s. Calculate the energy in its kinetic energy store.

$E_k = \frac{1}{2}mv^2 = \frac{1}{2} \times 2500 \times 20^2 = 500\ 000$ J

Raised Objects Store Energy in Gravitational Potential Energy Stores

1) <u>Lifting</u> an object in a <u>gravitational field</u> (page 88) requires <u>work</u>. This causes a <u>transfer of energy</u> to the <u>gravitational potential energy</u> (g.p.e.) store of the raised object. The <u>higher</u> the object is lifted, the <u>more</u> energy is transferred to this store.

2) The amount of energy in a gravitational potential energy store depends on the object's <u>mass</u>, its <u>height</u> and the <u>strength</u> of the gravitational field the object is in.

3) You can use this equation to find the <u>change in energy</u> in an object's gravitational potential energy store for a <u>change in height, h</u>.

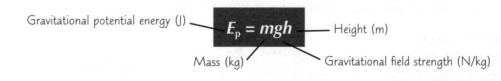

Gravitational potential energy (J) — $E_p = mgh$ — Height (m)

Mass (kg) Gravitational field strength (N/kg)

Stretching can Transfer Energy to Elastic Potential Energy Stores

<u>Stretching</u> or <u>squashing</u> an object can transfer energy to its <u>elastic potential energy store</u>. So long as the <u>limit of proportionality</u> has not been <u>exceeded</u> (page 94) energy in the <u>elastic potential energy store</u> of a stretched spring can be found using:

Spring constant (N/m)

Elastic potential energy (J) — $E_e = \frac{1}{2}ke^2$ — Extension (m)

Greater height means more energy in gravitational potential stores...

Wow, that's a lot of equations on a single page... As with all equations you come across, make sure you know what all the <u>variables</u> in them are, as well as what <u>units</u> all the variables in the equations are in.

Specific Heat Capacity

Specific heat capacity is really just a sciencey way of saying how hard it is to heat something up...

Different Materials Have Different Specific Heat Capacities

1) More energy needs to be transferred to the thermal energy store of some materials to increase their temperature than others.

2) For example:

> You need 4200 J to warm 1 kg of water by 1 °C, but only 139 J to warm 1 kg of mercury by 1 °C.

3) Materials that need to gain lots of energy in their thermal energy stores to warm up also transfer loads of energy when they cool down again. They can 'store' a lot of energy.

4) The measure of how much energy a substance can store is called its specific heat capacity.

> Specific heat capacity is the amount of energy needed to raise the temperature of 1 kg of a substance by 1 °C.

There's a Helpful Formula Involving Specific Heat Capacity

Below is the equation that links energy transferred to specific heat capacity (the Δ's just mean "change in").

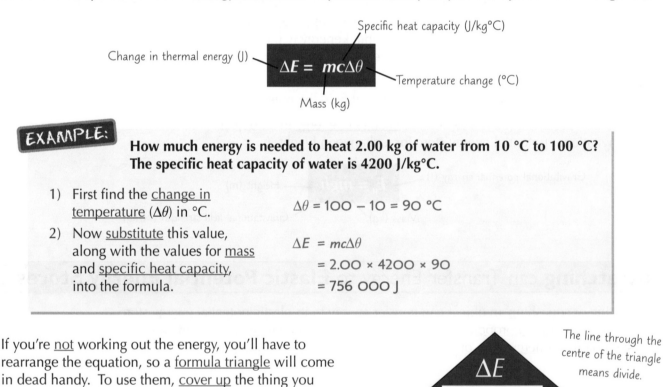

Specific heat capacity (J/kg°C)

Change in thermal energy (J)

$$\Delta E = mc\Delta\theta$$

Temperature change (°C)

Mass (kg)

EXAMPLE: **How much energy is needed to heat 2.00 kg of water from 10 °C to 100 °C? The specific heat capacity of water is 4200 J/kg°C.**

1) First find the change in temperature (Δθ) in °C.

$\Delta\theta = 100 - 10 = 90$ °C

2) Now substitute this value, along with the values for mass and specific heat capacity, into the formula.

$\Delta E = mc\Delta\theta$
$= 2.00 \times 4200 \times 90$
$= 756\ 000$ J

If you're not working out the energy, you'll have to rearrange the equation, so a formula triangle will come in dead handy. To use them, cover up the thing you want to find and write down what's left showing.
You write the bits of the formula in the triangle like this:

The line through the centre of the triangle means divide.

$$\frac{\Delta E}{m \times c \times \Delta\theta}$$

Some substances can store more energy than others...

Water is a substance that can store a lot of energy in its thermal stores — it has a high specific heat capacity. This is lucky for us as our bodies are mostly water. It'd be unfortunate if we started boiling on a hot day. Learn the definition of specific heat capacity and make sure you know how to use the formula involving it.

Investigating Specific Heat Capacity   PRACTICAL

This fun practical can be used to find out the specific heat capacity of a material.

You Can Investigate Specific Heat Capacities

1) To investigate a <u>solid</u> material (e.g. copper), you'll need a <u>block</u> of the material with <u>two holes</u> in it (for the <u>heater</u> and <u>thermometer</u> to go into, see the image on the right).

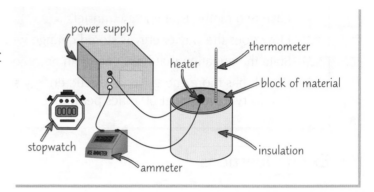

2) Measure the <u>mass</u> of the <u>block</u>, then wrap it in an insulating layer (e.g. a thick layer of newspaper) to <u>reduce</u> the <u>energy transferred</u> from the block to the <u>surroundings</u>. Insert the <u>thermometer</u> and <u>heater</u> as shown on the right.

3) Measure the <u>initial temperature</u> of the block and set the potential difference, V, of the power supply to be <u>10 V</u>. <u>Turn on</u> the power supply and <u>start</u> a <u>stopwatch</u>.

4) When you turn on the power, the <u>current</u> in the circuit (i.e. the moving charges) <u>does work</u> on the heater, transferring energy <u>electrically</u> from the power supply to the heater's <u>thermal energy store</u>. This energy is then transferred to the material's <u>thermal</u> energy store <u>by heating</u>, causing the material's <u>temperature</u> to increase.

5) As the block heats up, use the <u>thermometer</u> to measure its temperature e.g. <u>every minute</u>. Keep an eye on the <u>ammeter</u> — the <u>current</u> through the circuit, I, <u>shouldn't change</u>.

6) When you've collected enough readings (10 should do it), <u>turn off</u> the power supply.

7) Now you have to do some <u>calculations</u> to find the material's <u>specific heat capacity</u>:

- Using your measurement of the <u>current</u> and the <u>potential difference</u> of the <u>power supply</u>, you can calculate the <u>power</u> supplied to the heater, using <u>$P = VI$</u> (p.54). You can use this to calculate <u>how much energy</u>, E, has been <u>transferred to the heater</u> at the time of each temperature reading using the formula <u>$E = Pt$</u>, where t is the <u>time in seconds</u> since the experiment began.

- If you assume <u>all the energy</u> supplied to the heater has been <u>transferred to the block</u>, you can plot a <u>graph</u> of <u>energy transferred</u> to the thermal energy store of the block against <u>temperature</u>. It should look something like this: $\Rightarrow$

- Find the <u>gradient</u> of the straight part of the graph. This is $\Delta\theta \div \Delta E$. You know from the equation on the last page that $\Delta E = mc\Delta\theta$. So the specific heat capacity of the material of the block is: <u>$1 \div (\text{gradient} \times \text{the mass of the block})$</u>.

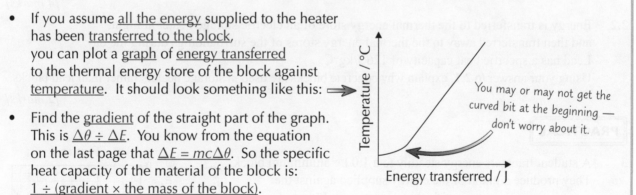

You may or may not get the curved bit at the beginning — don't worry about it.

8) You can <u>repeat</u> this experiment with <u>different materials</u> to see how their specific heat capacities <u>compare</u>.

You can also investigate the specific heat capacity of liquids — just place the heater and thermometer in an insulated beaker filled with a known mass of the liquid.

Think about how you could improve your experiments...

PRACTICAL TIP

If the hole in your material is <u>bigger</u> than your <u>thermometer</u>, you could put a small amount of water in the hole with the thermometer. This helps the thermometer to measure the temperature of the block more accurately, as water is a better thermal conductor than <u>air</u> (see page 24).

Warm-Up & Exam Questions

These questions give you chance to use your knowledge about energy transfers and specific heat capacity.

Warm-Up Questions

1) Give two methods of energy transfer.
2) How does the way energy is stored change when someone throws a ball upwards?
3) State the equation that links energy in an object's kinetic energy store with mass and speed.
4) Which has more energy in its kinetic energy store: a person walking at 3 miles per hour, or a lorry travelling at 60 miles per hour?

Exam Questions

1   A motor lifts a load of mass 20 kg.
     The load gains 137.2 J of energy in its gravitational potential energy store.

1.1   State the equation that links gravitational potential energy, mass, gravitational field strength and height.
      Use this equation to calculate the height through which the motor lifts the load.
      Assume the gravitational field strength = 9.8 N/kg

[4 marks]

1.2   The motor releases the load and the load falls.
      Ignoring air resistance, describe the changes in the way energy is stored that take place as the load falls.

[2 marks]

1.3   Describe how your answer to **1.2** would differ if air resistance was not ignored.

[1 mark]

2   36 000 J of energy is transferred to heat a 0.5 kg concrete block from 20 °C to 100 °C. (Grade 6-7)

2.1   Calculate the specific heat capacity of the concrete block.
      Use the correct equation from the Physics Equation Sheet on the inside back cover.

[4 marks]

2.2   Energy is transferred to the thermal energy store of an electric storage heater at night,
      and then transferred away to the thermal energy stores of the surroundings during the day.
      Lead has a specific heat capacity of 126 J/kg°C.
      Using your answer to **2.1**, explain why concrete blocks are used in storage heaters rather than lead blocks.

[2 marks]

PRACTICAL

3   A student transfers energy steadily to a 1.0 kg aluminium block.
     They produce a graph of the energy supplied against the
     increase in temperature of the block, shown in **Figure 1**.

3.1   Use **Figure 1** to find a value for the specific heat capacity
      of aluminium in J/kg°C. Use the correct equation from the
      Physics Equation Sheet on the inside back cover.

[4 marks]

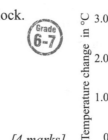

Figure 1

3.2   Would you expect the true value for the specific heat capacity of
      aluminium to be higher or lower than the value found in this experiment? Explain your answer.

[3 marks]

Conservation of Energy and Power

Repeat after me: energy is NEVER destroyed. Make sure you learn that fact, it's really important.

You Need to Know the Conservation of Energy Principle

1) The conservation of energy principle is that energy is always conserved:

> Energy can be transferred usefully, stored or dissipated, but can never be created or destroyed.

2) When energy is transferred between stores, not all of the energy is transferred usefully into the store that you want it to go to. Some energy is always dissipated when an energy transfer takes place.

3) Dissipated energy is sometimes called 'wasted energy' because the energy is being stored in a way that is not useful (usually energy has been transferred into thermal energy stores).

> A mobile phone is a system. When you use the phone, energy is usefully transferred from the chemical energy store of the battery in the phone. But some of this energy is dissipated in this transfer to the thermal energy store of the phone (you may have noticed your phone feels warm if you've been using it for a while).

4) You also need to be able to describe energy transfers for closed systems:

> A cold spoon is dropped into an insulated flask of hot soup, which is then sealed. You can assume that the flask is a perfect thermal insulator so the spoon and the soup form a closed system. Energy is transferred from the thermal energy store of the soup to the useless thermal energy store of the spoon (causing the soup to cool down slightly). Energy transfers have occurred within the system, but no energy has left the system — so the net change in energy is zero.

Power is the 'Rate of Doing Work' — i.e. How Much per Second

1) Power is the rate of energy transfer, or the rate of doing work.
2) Power is measured in watts. One watt = 1 joule of energy transferred per second.
3) You can calculate power using these equations:

Power (W) — $P = \dfrac{E}{t}$ — Energy transferred (J), Time (s)

Power (W) — $P = \dfrac{W}{t}$ — Work done (J), Time (s)

4) A powerful machine is not necessarily one which can exert a strong force (although it usually ends up that way). A powerful machine is one which transfers a lot of energy in a short space of time.

> Take two cars that are identical in every way apart from the power of their engines. Both cars race the same distance along a straight track to a finish line. The car with the more powerful engine will reach the finish line faster than the other car (it will transfer the same amount of energy but over less time).

EXAMPLE:

It takes 8000 J of work to lift a stunt performer to the top of a building. Motor A can lift the stunt performer to the correct height in 50 s. Motor B would take 300 s to lift the performer to the same height. Which motor is most powerful? Calculate the power of this motor.

1) Both motors transfer the same amount of energy, but motor A would do it quicker than motor B.

So motor A is the most powerful motor.

2) Plug the time taken and work done for motor A into the equation $P = W \div t$ and find the power.

$P = W \div t$
$= 8000 \div 50 = 160$ W

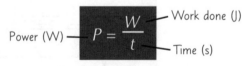

Conduction

You're probably familiar with the idea that metals are <u>good thermal conductors</u> but wood isn't, and so on. This page is about how conduction <u>actually happens</u> and about the <u>energy transfers</u> that take place.

Conduction Occurs Mainly in Solids

<u>Conduction</u> is the process where <u>vibrating particles</u> <u>transfer energy</u> to <u>neighbouring particles</u>.

1) Energy transferred to an object <u>by heating</u> is transferred to the <u>thermal store</u> of the object. This energy is shared across the <u>kinetic</u> energy stores of the <u>particles</u> in the object.

2) The particles in the part of the object being heated <u>vibrate</u> more and <u>collide</u> with each other. These <u>collisions</u> cause energy to be transferred between particles' <u>kinetic</u> energy stores. This is <u>conduction</u>.

H O T   ENERGY TRANSFER   C O L D

3) This process <u>continues throughout</u> the object until the energy is transferred to the <u>other side</u> of the object. It's then usually transferred to the <u>thermal</u> energy store of the <u>surroundings</u> (or anything else <u>touching</u> the object).

Particles in liquids and gases are much more free to move around, which is why they usually transfer energy by convection (see the next page) instead of conduction.

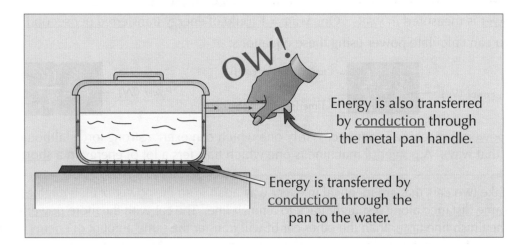

OW!

Energy is also transferred by <u>conduction</u> through the metal pan handle.

Energy is transferred by <u>conduction</u> through the pan to the water.

4) <u>Thermal conductivity</u> is a measure of how <u>quickly</u> energy is transferred through a material in this way. Materials with a <u>high thermal conductivity</u> transfer energy between their particles <u>quickly</u>.

Some substances are better thermal conductors than others...

<u>Denser</u> materials (see page 64) are usually <u>better conductors</u> than less dense materials. It's easy to see why — particles that are <u>right next to</u> each other will pass energy between their <u>kinetic energy stores</u> far more effectively than particles that are far apart. For example, water is a much better thermal conductor than air.

Convection

You'll have heard it said that <u>hot air rises</u>. Well, it <u>does</u> (assuming the air above it is cooler).

Convection Occurs Only in **Liquids** and **Gases**

> <u>Convection</u> is where energetic particles
> <u>move away</u> from <u>hotter</u> to <u>cooler regions</u>.

1) <u>Convection</u> can happen in <u>gases</u> and <u>liquids</u>. Energy is transferred <u>by heating</u> to the <u>thermal store</u> of the liquid or gas. As with conduction, this energy is shared across the <u>kinetic</u> energy stores of the gas or liquid's particles.

2) Unlike in solids, the particles in liquids and gases are <u>able to move</u>. When you heat a region of a gas or liquid, the particles <u>move faster</u> and the <u>space</u> between individual particles <u>increases</u>. This causes the <u>density</u> (p.64) of the <u>region</u> being heated to <u>decrease</u>.

3) Because liquids and gases can <u>flow</u>, the warmer and less dense region will <u>rise</u> above <u>denser</u>, <u>cooler</u> regions. If there is a <u>constant</u> heat source, a <u>convection current</u> can be created.

Radiators Create **Convection Currents**

1) Heating a room with a <u>radiator</u> relies on creating <u>convection currents</u> in the <u>air</u> of the room.

2) Energy is <u>transferred from</u> the <u>radiator</u> to the nearby <u>air particles</u> by <u>conduction</u> (the air particles collide with the radiator surface).

3) The air by the radiator becomes <u>warmer</u> and <u>less dense</u> (as the particles move <u>quicker</u>).

4) This <u>warm air rises</u> and is replaced by <u>cooler air</u>. The cooler air is then heated by the radiator.

5) At the same time, the previously heated air transfers energy to the surroundings (e.g. the walls and contents of the room). It <u>cools</u>, becomes <u>denser</u> and <u>sinks</u>.

6) This cycle <u>repeats</u>, causing a <u>flow of air</u> to circulate around the room — this is a <u>convection current</u>.

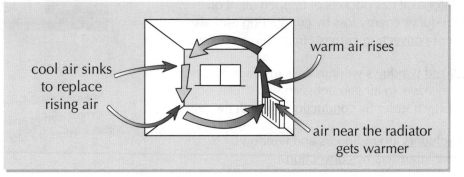

cool air sinks to replace rising air

warm air rises

air near the radiator gets warmer

In convection, particles move from hotter regions to cooler regions...

So in <u>convection</u>, the particles move taking their energy with them. Don't get this confused with <u>conduction</u> though. Conduction is the process where <u>vibrating particles</u> transfer energy to neighbouring particles. Have a flick back to the last page if you need to remind yourself about this.

Reducing Unwanted Energy Transfers

There are a few ways you can <u>reduce</u> the amount of energy scampering off to a <u>completely useless</u> store — <u>lubrication</u> and <u>thermal insulation</u> are the ones you need to know about. Read on to find out more...

Lubrication Reduces **Frictional Forces**

1) Whenever something <u>moves</u>, there's usually at least one <u>frictional force</u> acting against it (p.108). This causes some energy in the system to be <u>dissipated</u> (p.23), e.g. <u>air resistance</u> can transfer energy from a falling object's <u>kinetic energy store</u> to its <u>thermal energy store</u>.

2) For objects that are being rubbed together, <u>lubricants</u> can be used to reduce the friction between the objects' surfaces when they move. Lubricants are usually <u>liquids</u> (like <u>oil</u>), so they can <u>flow</u> easily between objects and <u>coat</u> them.

Streamlining reduces air resistance too, see page 108.

Insulation Reduces the Rate of **Energy Transfer** by **Heating**

The last thing you want when you've made your house nice and toasty is for that energy to <u>escape</u> outside. There are a few things you can do to <u>prevent energy losses</u> through <u>heating</u>:

* Have <u>thick walls</u> that are made from a material with a <u>low thermal conductivity</u>. The <u>thicker</u> the walls and the <u>lower</u> their <u>thermal conductivity</u>, the <u>slower</u> the rate of energy transfer will be (so the building will <u>cool more slowly</u>).

* Use <u>thermal insulation</u>. Here are some examples:

1) Some houses have <u>cavity walls</u>, made up of an <u>inner</u> and an <u>outer</u> wall with an air gap in the middle. The <u>air gap</u> reduces the amount of energy transferred by conduction through the walls. <u>Cavity wall insulation</u>, where the cavity wall air gap is filled with a <u>foam</u>, can also reduce energy transfer by <u>convection</u> in the wall cavity.

2) <u>Loft insulation</u> can be laid out across the loft floor and ceiling. Fibreglass wool is often used which is a <u>good insulator</u> as it has pockets of trapped air. Loft insulation reduces energy loss by <u>conduction</u> and also helps prevent <u>convection</u> currents from being created.

3) <u>Double-glazed windows</u> work in the same way as cavity walls — they have an air gap between two sheets of glass to prevent energy transfer by <u>conduction</u> through the windows.

4) <u>Draught excluders</u> around doors and windows reduce energy transfers by <u>convection</u>.

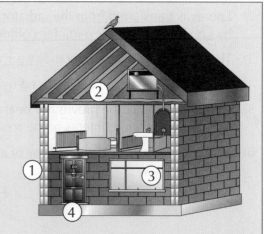

Reducing the difference between the temperature inside and outside the house will also reduce the rate of energy transfer.

Having a well-insulated house can reduce your heating bills...

When people talk of <u>energy loss</u>, it's <u>not</u> that the energy has disappeared. It still exists (see page 23), just not necessarily in the <u>store</u> we want. For example, in a car, you want the energy to transfer to the <u>kinetic energy store</u> of the wheels, and not to the <u>thermal energy stores</u> of the moving components.

Investigating Energy Transfers PRACTICAL

Now it's time to get practical. This investigation isn't too complicated, but there are several variables that it's really important to keep constant. So make sure you follow the method carefully.

You can **Investigate** the **Effectiveness of Different Insulators**...

Here's the method for how you can do this:

1) Boil water in a kettle. Pour some of the water into a sealable container (e.g. a beaker and lid) to a safe level. Measure the mass of water in the container (p.182).

2) Use a thermometer to measure the initial temperature of the water (p.182).

3) Seal the container and leave it for five minutes. Measure this time using a stopwatch.

4) Remove the lid and measure the final temperature of the water.

5) Pour away the water and allow the container to cool to room temperature.

6) Repeat this experiment, but wrap the container in a different material (e.g. foil, newspaper) once it has been sealed. Make sure you use the same mass of water each time.

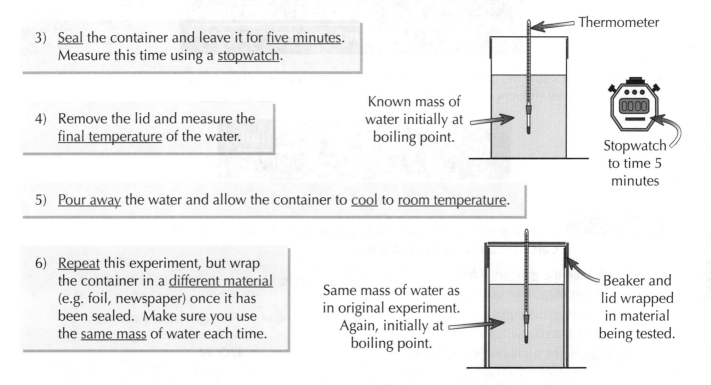

Thermometer

Known mass of water initially at boiling point.

Stopwatch to time 5 minutes

Same mass of water as in original experiment. Again, initially at boiling point.

Beaker and lid wrapped in material being tested.

You should find that the temperature difference (and so the energy transferred) is reduced by wrapping the container in thermally insulating materials like bubble wrap or newspaper.

... and **Factors** that Affect how **Good** a Material is at **Insulating**

1) You could, for example, also investigate how the thickness of the material affects the temperature change of the water.

2) You should find that the thicker the insulating layer, the less energy is transferred and the smaller the temperature change of the water.

It's tricky to keep the variables constant in this experiment...

For example, it's unlikely that the foil you have available will be exactly the same thickness as newspaper. These are all things you should mention in your conclusion. It's a good idea to repeat your investigation too — you don't want errors creeping in without you noticing.

Efficiency

Devices have energy transferred to them, but only transfer some of that energy to useful energy stores. Wouldn't it be great if we could tell how much it usefully transfers? That's where efficiency comes in.

Most **Energy Transfers** Involve Some **Waste Energy**

1) Useful devices are only useful because they can transfer energy from one store to another.

2) As you'll probably have gathered by now, some of the input energy is usually wasted by being transferred to a useless energy store — usually a thermal energy store.

3) The less energy that is 'wasted' in this energy store, the more efficient the device is said to be.

4) You can improve the efficiency of energy transfers by insulating objects, lubricating them or making them more streamlined (see pages 26 and 108).

5) The efficiency for any energy transfer can be worked out using this equation:

$$\text{Efficiency} = \frac{\text{Useful output energy transfer}}{\text{Total input energy transfer}}$$

You can give efficiency as a decimal or you can multiply your answer by 100 to get a percentage, i.e. 0.75 or 75%.

6) You might not know the energy inputs and outputs of a device, but you can still calculate its efficiency as long as you know the power input and output:

$$\text{Efficiency} = \frac{\text{Useful power output}}{\text{Total power input}}$$

EXAMPLE: **A blender is 70% efficient. It has a total input power of 600 W. Calculate the useful output power.**

1) Change the efficiency from a percentage to a decimal.

efficiency = 70% = 0.70

2) Rearrange the equation for useful power output.

useful power output = efficiency × total power input

3) Stick in the numbers you're given.

= 0.70 × 600
= 420 W

Useful Energy Output Isn't Usually Equal to **Total Energy Input**

1) For any given example you can talk about the types of energy being input and output, but remember — NO device is 100% efficient and the wasted energy is usually transferred to useless thermal energy stores.

2) Electric heaters are the exception to this. They're usually 100% efficient because all the energy in the electrostatic energy store is transferred to "useful" thermal energy stores.

3) Ultimately, all energy ends up transferred to thermal energy stores. For example, if you use an electric drill, its energy is transferred to lots of different energy stores, but quickly ends up all in thermal energy stores.

Warm-Up & Exam Questions

Don't let your energy dissipate. These questions will let you see how efficient your revision has been.

Warm-Up Questions

1) State the principle of the conservation of energy.
2) What is power? State the units it is measured in.
3) Name two mechanisms in which energy is transferred by heating.
4) Give one way you could reduce the frictional forces in the hinge of an automatic door?
5) For a given material, how does its thermal conductivity affect the rate of energy transfer through it?
6) Why is the efficiency of an appliance always less than 100%?

Exam Questions

1 The motor of an electric scooter moves the scooter 10 metres along a flat, horizontal **(Grade 4-6)** course in 20 seconds. During this time the motor does 1000 J of work.

1.1 Write down the equation that links power, work done and time.
Use this equation to calculate the power of the motor.

[3 marks]

1.2 The moving parts of the scooter are lubricated. The scooter then completes the course in 18 seconds.
Explain, in terms of energy transfer, why the scooter completes the course in a faster time.

[2 marks]

1.3 The scooter's motor is replaced with a more powerful, but otherwise identical, motor.
It moves along the same 10 m course.
Describe how its performance will differ from before. Explain your answer.

[2 marks]

PRACTICAL

2 A student wants to test whether the temperature difference between the inside and outside of a flask affects the rate of energy transfer **(Grade 6-7)** from the flask. She sets up the experiment shown in **Figure 1** and records the temperature of the water in each flask every minute.

Figure 1

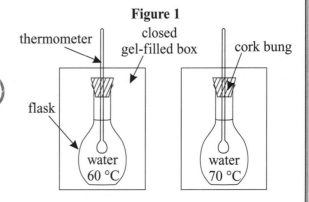

2.1 Give **two** variables that the student must control.

[2 marks]

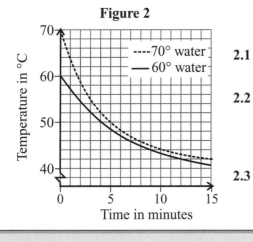

Figure 2

2.2 The student produces the graph in **Figure 2** of her results.
Which temperature of water shows the greatest initial rate of energy transfer? Explain your answer using the graph.

[2 marks]

2.3 Calculate the rate of temperature change after 5 minutes for the 60 °C water. Give your answer in °C / minute.

[3 marks]

Exam Questions

3   Each second, a 1000 W hairdryer transfers 280 J of energy away
     through sound waves, 200 J of energy to kinetic energy stores,
     and the rest directly to the thermal energy store of the heating element.

Grade 6-7

3.1   Calculate how much energy the hairdryer transfers directly to thermal energy stores each second.

[1 mark]

3.2   Calculate the efficiency of the hairdryer.

[3 marks]

3.3   It takes 4 minutes for a particular person to dry their hair.
      Calculate how much energy from electrical stores, in joules, is transferred in this time.

[3 marks]

4   Torch A transfers 1200 J of energy per minute.
     480 J of this is transferred away usefully as light, 690 J is transferred
     to useless thermal energy stores and 30 J is transferred away as sound.

Grade 6-7

4.1   Write down the equation linking efficiency, useful output energy transfer and total input energy transfer.

[1 mark]

4.2   Calculate the efficiency of torch A.

[2 marks]

4.3   Torch B transfers 600 J of energy away usefully by light each minute.
      Calculate the output power of torch B.

[2 marks]

4.4   Torch B has an efficiency of 0.55. Calculate input power of torch B.

[3 marks]

4.5   Each torch is powered by an identical battery. A student claims that the battery in torch B will
      go 'flat' quicker than in torch A because it transfers more energy away as light each minute.
      Explain whether or not you agree with the student.

[2 marks]

PRACTICAL

Figure 3

5   A scientist is worried that her house is losing
     a lot of energy through its walls and windows.

Grade 7-9

5.1   Suggest **one** way that the scientist could reduce
      the energy lost through the walls of her house.

[1 mark]

      A shop sells two different types of transparent
      film that both claim to substantially
      'reduce energy loss' when stuck over windows.

transparent films

beakers

polystyrene lids

thermometers

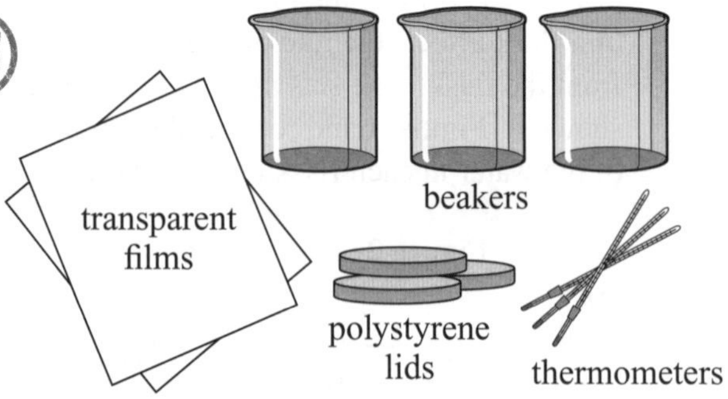

5.2*   Describe a method the scientist could use, using the apparatus in **Figure 3**,
       to investigate whether the claims are true and which type of film is best.

[6 marks]

5.3   The scientist is concerned that her results are not reproducible.
      Suggest **one** way she could check if her results are reproducible.

[1 mark]

Energy Resources and their Uses

There are lots of <u>energy resources</u> available on Earth. They are either <u>renewable</u> or <u>non-renewable</u> resources.

Non-Renewable Energy Resources Will **Run Out** One Day

<u>Non-renewable</u> energy resources are <u>fossil fuels</u> and <u>nuclear fuel</u> (e.g. uranium and plutonium). <u>Fossil fuels</u> are natural resources that form <u>underground</u> over <u>millions</u> of years. They are typically <u>burnt</u> to provide energy. The <u>three main</u> fossil fuels are:

1) Coal

2) Oil

3) (Natural) Gas

- These will <u>all</u> 'run out' one day.
- They all do <u>damage</u> to the environment.
- But they provide <u>most of our energy</u>.

Renewable Energy Resources Will **Never** Run Out

<u>Renewable</u> energy resources are:

1) The Sun (Solar)

2) Wind

3) Water waves

4) Hydro-electricity

5) Bio-fuel

6) Tides

7) Geothermal

- These will <u>never run out</u> — the energy can be 'renewed' as it is used.
- Most of them do <u>damage</u> the environment, but in <u>less nasty</u> ways than non-renewables.
- The trouble is they <u>don't</u> provide much <u>energy</u> and some of them are <u>unreliable</u> because they depend on the weather.

Energy Resources can be Used for **Transport**...

<u>Transport</u> is one of the most obvious places where <u>fuel</u> is used. Here are a few transportation methods that use either <u>renewable</u> or <u>non-renewable</u> energy resources:

Electricity can also be used to power vehicles, (e.g. trains and some cars). It can be generated using renewable or non-renewable energy resources (p.32-36).

NON-RENEWABLE ENERGY RESOURCES
- <u>Petrol</u> and <u>diesel</u> powered vehicles (including most cars) use fuel created from <u>oil</u>.
- <u>Coal</u> is used in some old-fashioned <u>steam trains</u> to boil water to produce steam.

RENEWABLE ENERGY RESOURCES
Vehicles that run on pure <u>bio-fuels</u> (p.35) or a <u>mix</u> of a bio-fuel and petrol or diesel (only the bio-fuel bit is renewable, though).

...And for **Heating**

<u>Energy resources</u> are also needed for <u>heating</u> things like your home.

NON-RENEWABLE ENERGY RESOURCES
- <u>Natural gas</u> is the most widely used fuel for heating homes in the UK. The gas is used to heat <u>water</u>, which is then pumped into <u>radiators</u> throughout the home.
- <u>Coal</u> is commonly burnt in fireplaces.
- <u>Electric heaters</u> (sometimes called storage heaters) which use electricity generated from <u>non-renewable</u> energy resources.

RENEWABLE ENERGY RESOURCES
- A <u>geothermal</u> (or ground source) <u>heat pump</u> uses geothermal energy resources (p.33) to heat buildings.
- <u>Solar water heaters</u> work by using the sun to heat <u>water</u> which is then pumped into radiators in the building.
- Burning <u>bio-fuel</u> or using <u>electricity</u> generated from renewable resources can also be used for heating.

Wind and Solar Power

Renewable energy resources, like wind and solar resources, will not run out. They don't generate as much electricity as non-renewables though — if they did we'd all be using solar-powered toasters by now.

Wind Power — Lots of Little Wind Turbines

This involves putting lots of wind turbines (windmills) up in exposed places like on moors or round coasts.

1) Each turbine has a generator inside it — the rotating blades turn the generator and produce electricity.

2) There's no pollution (except for a bit when they're manufactured).

3) But they do spoil the view. You need about 1500 wind turbines to replace one coal-fired power station and 1500 of them cover a lot of ground — which would have a big effect on the scenery.

4) And they can be very noisy, which can be annoying for people living nearby.

5) There's also the problem of the turbines stopping when the wind stops or if the wind is too strong, and it's impossible to increase supply when there's extra demand (p.55). On average, wind turbines produce electricity 70-85% of the time.

6) The initial costs are quite high, but there are no fuel costs and minimal running costs.

7) There's no permanent damage to the landscape — if you remove the turbines, you remove the noise and the view returns to normal.

Solar Cells — Expensive but Not Much Environmental Damage

Solar cells generate electric currents directly from sunlight.

1) Solar cells are often the best source of energy to charge batteries in calculators and watches which don't use much electricity.

2) Solar power is often used in remote places where there's not much choice (e.g. the Australian outback) and to power electric road signs and satellites.

3) There's no pollution. (Although the factories do use quite a lot of energy and produce some pollution when they manufacture the cells.)

4) In sunny countries solar power is a very reliable source of energy — but only in the daytime. Solar power can still be cost-effective in cloudy countries like Britain though.

5) Like wind, you can't increase the power output when there is extra demand.

6) Initial costs are high but after that the energy is free and running costs almost nil.

7) Solar cells are usually used to generate electricity on a relatively small scale.

electric current

sunlight

solar cell

electrical components

People love the idea of wind power — just not in their back yard...

It's easy to think that non-renewables are the answer to the world's energy problems. However, they have their downsides, and we definitely couldn't rely on them totally at present. Make sure you know the pros and cons for wind and solar power because there are more renewables coming up on the next page.

Geothermal and Hydro-electric Power

Here are some more examples of <u>renewable energy resources</u> — <u>geothermal</u> and <u>hydro-electric</u>. These ones are a bit more <u>reliable</u> than wind and solar — read on to find out why.

Geothermal Power — Energy from **Underground**

Geothermal power uses energy from <u>underground thermal energy</u> stores.

1) This is <u>only possible</u> in <u>volcanic areas</u> where <u>hot rocks</u> lie quite near to the <u>surface</u>. The source of much of the energy is the <u>slow decay</u> of various <u>radioactive elements</u>, including <u>uranium</u>, deep inside the Earth.

2) This is actually <u>brilliant free energy</u> that's <u>reliable</u> with very few environmental problems.

3) Geothermal energy can be used to <u>generate electricity</u>, or to <u>heat buildings directly</u>.

4) The <u>main drawbacks</u> with geothermal energy are that there <u>aren't</u> very many <u>suitable locations</u> for power plants and that the <u>cost</u> of building a power plant is often <u>high</u> compared to the <u>amount</u> of energy it produces.

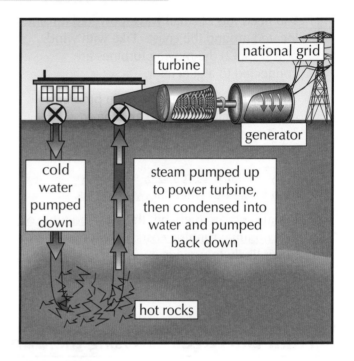

Hydro-electric Power Uses **Falling Water**

Hydro-electric power transfers energy from the <u>kinetic store</u> of <u>falling water</u>.

1) <u>Hydro-electric power</u> usually requires the <u>flooding</u> of a valley by building a big <u>dam</u>. <u>Rainwater</u> is caught and allowed out <u>through turbines</u>. There is <u>no pollution</u> (as such).

2) But there is a <u>big impact</u> on the <u>environment</u> due to the flooding of the valley (rotting vegetation releases methane and carbon dioxide) and possible <u>loss of habitat</u> for some species (sometimes the loss of whole villages). The reservoirs can also look very <u>unsightly</u> when they <u>dry up</u>. Putting hydro-electric power stations in <u>remote valleys</u> tends to reduce their impact on <u>humans</u>.

3) A <u>big advantage</u> is it can provide an <u>immediate response</u> to an increased demand for electricity.

4) There's no problem with <u>reliability</u> except in times of <u>drought</u> — but remember this is Great Britain we're talking about.

5) <u>Initial costs</u> are <u>high</u>, but there are <u>no fuel costs</u> and <u>minimal running costs</u>.

6) It can be a useful way to generate electricity on a <u>small scale</u> in <u>remote areas</u>.

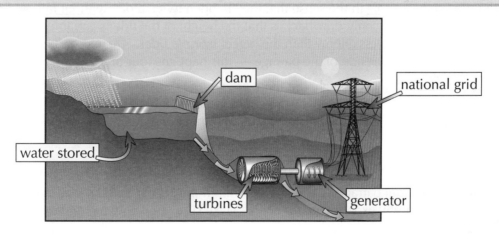

Wave Power and Tidal Barrages

Good ol' water. Not only can we drink it, we can also use it to generate electricity. It's easy to get confused between wave and tidal power as they both involve the seaside — but don't. They are completely different.

Wave Power — Lots of Little Wave-Powered Turbines

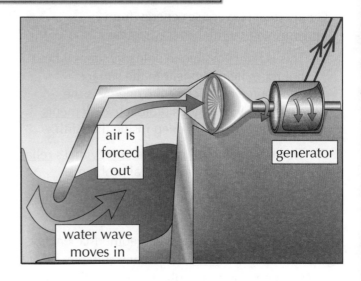

1) You need lots of small wave-powered turbines located around the coast. Like with wind power (p.32) the moving turbines are connected to a generator.

2) There is no pollution. The main problems are disturbing the seabed and the habitats of marine animals, spoiling the view and being a hazard to boats.

3) They are fairly unreliable, since waves tend to die out when the wind drops.

4) Initial costs are high, but there are no fuel costs and minimal running costs. Wave power is never likely to provide energy on a large scale, but it can be very useful on small islands.

Tidal Barrages — Using the Sun and Moon's Gravity

1) Tides are used in lots of ways to generate electricity. The most common method is building a tidal barrage.

2) Tidal barrages are big dams built across river estuaries, with turbines in them. As the tide comes in it fills up the estuary. The water is then allowed out through turbines at a controlled speed.

3) Tides are produced by the gravitational pull of the Sun and Moon.

4) There is no pollution. The main problems are preventing free access by boats, spoiling the view and altering the habitat of the wildlife, e.g. wading birds and sea creatures who live in the sand.

5) Tides are pretty reliable in the sense that they happen twice a day without fail, and always near to the predicted height. The only drawback is that the height of the tide is variable so lower (neap) tides will provide significantly less energy than the bigger (spring) tides. They also don't work when the water level is the same either side of the barrage — this happens four times a day because of the tides.

6) Initial costs are moderately high, but there are no fuel costs and minimal running costs. Even though it can only be used in some of the most suitable estuaries tidal power has the potential for generating a significant amount of energy.

Wave and tidal — power from the motion of the ocean...

The first large-scale tidal barrages started being built in the 1960s, so tidal power isn't a new thing. Wave power is still pretty experimental though. Make sure you know the differences in how they work.

Bio-fuels

And the energy resources just keep on coming. It's over soon, I promise. Just a few more to go.

Bio-fuels are Made from **Plants** and **Waste**

Bio-fuels are renewable energy resources created from either plant products
or animal dung. They can be solid, liquid or gas and can be burnt to
produce electricity or run cars in the same way as fossil fuels.

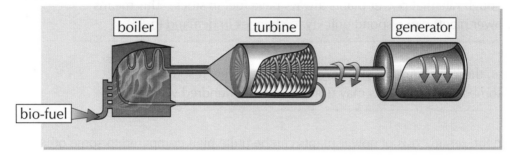

They have **Pros...**

1) They are supposedly carbon neutral, although there is some debate about this as it's
only really true if you keep growing plants at the rate that you're burning things.

2) Bio-fuels are fairly reliable, as crops take a relatively short time to grow
and different crops can be grown all year round. However, they cannot
respond to immediate energy demands. To combat this, bio-fuels are
continuously produced and stored for when they are needed.

... and **Cons**

1) The cost to refine bio-fuels so they are suitable for use is very high.

2) Some people worry that growing crops specifically for bio-fuels will mean there
isn't enough space or water to meet the demands for crops that are grown for food.

3) In some regions, large areas of forest have been cleared to make room to grow
bio-fuels, resulting in lots of species losing their natural habitats. The decay and
burning of this vegetation also increases carbon dioxide (CO_2) and methane emissions.

In theory, bio-fuels are carbon neutral...

Stuff you've learnt in biology may help you get your head around this one. When plants grow,
they absorb CO_2 from the atmosphere for photosynthesis, but when you burn bio-fuels, you
release CO_2 into the atmosphere. A bio-fuel is 'carbon neutral' if the amount of CO_2 released by
burning it is equal to the amount absorbed by the plants you grow to make the bio-fuel.

Non-Renewable Resources

Renewable resources may sound like great news for the environment. But when it comes down to it, they don't currently meet all our needs so we still need those nasty, polluting non-renewables.

Non-Renewables are Reliable...

1) Fossil fuels and nuclear energy are reliable. There's enough fossil and nuclear fuels to meet current demand, and they are extracted from the Earth at a fast enough rate that power plants always have fuel in stock. This means that the power plants can respond quickly to changes in demand (p.55).

Nuclear power plants use fission to produce electricity (p.84).

2) However, these fuels are slowly running out. If no new resources are found, some fossil fuel stocks may run out within a hundred years.

3) While the set-up costs of power plants can be quite high compared to some other energy resources, the running costs aren't that expensive. Combined with fairly low fuel extraction costs, using fossil fuels is a cost effective way to produce energy (which is why it's so popular).

...But Create Environmental Problems

1) Coal, oil and gas release carbon dioxide (CO_2) into the atmosphere when they're burned. All this CO_2 adds to the greenhouse effect, and contributes to global warming.

2) Burning coal and oil also releases sulfur dioxide, which causes acid rain — which can be harmful to trees and soils and can have far-reaching effects in ecosystems.

3) Acid rain can be reduced by taking the sulfur out before the fuel is burned, or cleaning up the emissions.

4) Views can be spoilt by fossil fuel power plants, and coal mining makes a mess of the landscape, especially "open-cast mining".

5) Oil spillages cause serious environmental problems, affecting mammals and birds that live in and around the sea. We try to avoid them, but they'll always happen.

Radiation can be very dangerous to humans — see p.83 for more.

6) Nuclear power is clean but the nuclear waste is very dangerous and difficult to dispose of.

7) Nuclear fuel (e.g. uranium or plutonium) is relatively cheap but the overall cost of nuclear power is high due to the cost of the power plant and final decommissioning.

8) Nuclear power always carries the risk of a major catastrophe like the Fukushima disaster in Japan.

Currently we Depend on Fossil Fuels

1) Over the 20th century, the electricity use of the UK hugely increased as the population grew and people began to use electricity for more and more things.

2) Since the beginning of the 21st century, electricity use in the UK has been decreasing (slowly), as we get better at making appliances more efficient (p.28) and become more careful with energy use in our homes.

3) Most of our electricity is produced using fossil fuels (mostly coal and gas) and from nuclear power.

4) Generating electricity isn't the only reason we burn fossil fuels — oil (diesel and petrol) is used to fuel cars, and gas is used to heat homes and cook food.

Trends in Energy Resource Use

Non-renewables may be what we rely on for the vast majority of our energy needs at the moment.
But the balance may soon start shifting...

The Aim is to **Increase Renewable Energy** Use

We are trying to increase our use of renewable energy resources (the UK aims to
use renewable resources to provide 15% of its total yearly energy by 2020).
This move towards renewable energy resources has been triggered by many things:

1) We now know that burning fossil fuels is very damaging to the environment (see last page). This
   makes many people want to use more renewable energy resources that affect the environment less.

2) People and governments are also becoming increasingly aware that
   non-renewables will run out one day. Many people think it's better
   to learn to get by without non-renewables before this happens.

3) Pressure from other countries and the public has meant that governments have begun to introduce
   targets for using renewable resources. This in turn puts pressure on energy providers to build new
   power plants that use renewable resources to make sure they do not lose business and money.

4) Car companies have also been affected by this change in attitude towards the
   environment. Electric cars and hybrids (cars powered by two fuels, e.g. petrol
   and electricity) are already on the market and their popularity is increasing.

It's **Not** That **Straightforward** Though

The use of renewables is limited by reliability, money and politics.

1) There's lots of scientific evidence supporting renewables, but although scientists can give advice,
   they don't have the power to make people, companies or governments change their behaviour.

2) Building new renewable power plants costs money, so some energy providers are reluctant to do this,
   especially when fossil fuels are so cost effective. The cost of switching to renewable power will have
   to be paid, either by customers in their bills, or through government and taxes. Some people don't
   want to or can't afford to pay, and there are arguments about whether it's ethical to make them.

3) Even if new power plants are built, there are arguments over where to put them. E.g. many people
   don't want to live next to a wind farm, causing protests. There are arguments over whether it's ethical
   to make people put up with wind farms built next to them when they may not agree with them.

4) Some energy resources like wind power are not as reliable as traditional fossil fuels, whilst others
   cannot increase their power output on demand. This would mean either having to use a combination
   of different power plants (which would be expensive) or researching ways to improve reliability.

5) Research on improving the reliability and cost of renewables takes time and money — it may be years
   before improvements are made, even with funding. Until then, we need non-renewable power.

6) Making personal changes can also be quite expensive. Hybrid cars are generally more expensive
   than equivalent petrol cars and things like solar panels for your home are still quite pricey.
   The cost of these things is slowly going down, but they are still not an option for many people.

Warm-Up & Exam Questions

This is the last set of warm-up and exam questions on Topic 1. They're not *too* horrendous, I promise.

Warm-Up Questions

1) Name three non-renewable energy resources.
2) Give one advantage and one disadvantage associated with the reliability of renewable resources.
3) Describe one way that renewable energy resources can be used to power vehicles.
4) Give two ways in which using coal as an energy resource causes environmental problems.
5) Suggest two reasons why we can't just stop using fossil fuels immediately.

Exam Questions

1   The inhabitants of a remote island do not have the resources or expertise   (Grade 4-6) to build a nuclear power plant. They have no access to fossil fuels.

1.1   The islanders have considered using wind, solar and hydro-electric power to generate electricity. Suggest **two** other renewable energy resources they could use.

[2 marks]

1.2   The islanders decide that hydro-electric power could reliably generate enough electricity for all their needs, but they are concerned about the environmental impact. Give **one** environmental impact of using hydro-electric power to generate electricity.

[1 mark]

2   In the hydro-electric power station in **Figure 1**, water is held back behind a dam before (Grade 4-6) being allowed to flow out through turbines.

Figure 1

2.1   Describe the transfer between energy stores of the water which occurs during this process.

[2 marks]

2.2   The tides can also be used to generate electricity using tidal barrages. Give **two** environmental advantages of generating electricity using tidal barrages.

[2 marks]

3   A family want to install solar panels on their roof. They have 10 m² of space on their roof for the solar panels. They use 32 500 000 J of energy per day. (Grade 6-7) A 1 m² solar panel has an output of 200 W in good sunlight.

3.1   Calculate the minimum number of 1 m² solar panels required to cover the family's daily energy use, assuming there are 5 hours of good sunlight in a day.

[5 marks]

3.2   Determine, using your answer from **3.1**, whether the family can install enough solar panels to provide all of the energy they use, assuming there are 5 hours of good sunlight every day.

[1 mark]

3.3   In reality, the number of hours of good sunlight in a day varies based on the weather and time of year. Discuss the reliability of energy from solar panels compared to from a local coal-fired power station.

[3 marks]

Revision Summary for Topic 1

Well, that's that for <u>Topic 1</u> — this is when you find out <u>how much of it went in</u>.
- Try these questions and <u>tick off each one</u> when you <u>get it right</u>.
- When you've done <u>all the questions</u> under a heading and are <u>completely happy</u> with it, tick it off.

Energy Stores and Systems (p.17-19) ☑

1) Write down four energy stores.
2) What is a system?
3) Describe the energy transfers that occur when a car collides with a stationary object.
4) If energy is transferred to an object's kinetic energy store, what happens to its speed?
5) Give the equation for finding the energy in an object's gravitational potential energy store.
6) What kind of energy store is energy transferred to when you compress a spring?
7) What does the variable 'e' stand for in the equation for energy in an elastic potential energy store?

Specific Heat Capacity (p.20-21) ☑

8) What is the definition of the specific heat capacity of a material?
9) Suggest why a material with a high specific heat capacity is better suited
   for use in a heating system than a material with a low specific heat capacity.
10) Describe an experiment to find the specific heat capacity of a material.

Conservation of Energy and Power (p.23) ☑

11) True of false? Energy can be destroyed.
12) Give two equations to calculate power.
13) How much energy is transferred each second to a 50 W device?

Reducing Unwanted Energy Transfers and Improving Efficiency (p.24-28) ☑

14) True or false? A high thermal conductivity means there is a high rate of energy transfer.
15) How can you reduce unwanted energy transfers in a machine with moving components?
16) Give four ways to prevent unwanted energy transfers in a home.
17) True or false? Thicker walls make a house cool down quicker.
18) Describe an experiment you could do to investigate ways of reducing unwanted energy transfers.
19) What is the efficiency of an energy transfer? Give the equation that relates efficiency to power.

Energy Resources and Trends in their Use (p.31-37) ☑

20) Name four renewable energy resources.
21) What is the difference between renewable and non-renewable energy resources?
22) Give an example of how a renewable energy resource is used in everyday life.
23) Explain why solar power is considered to be a fairly reliable energy resource.
24) Give one environmental impact of using wave power to generate electricity.
25) Describe how you can reduce the acid rain caused by burning coal and oil.
26) Give one environmental benefit of using nuclear power.
27) Explain why the UK plans to use more renewable energy resources in the future.

Current and Circuit Symbols

Isn't <u>electricity</u> great? Mind you it's pretty bad news if the <u>words</u> don't mean anything to you...

Current is the flow of Electric Charge

1) <u>Electric current</u> is the <u>flow</u> of electric charge round the circuit.
Current will <u>only flow</u> around a complete (closed) circuit if there's
a <u>potential difference</u>. So a current can only flow if there's a
source of potential difference. Unit of current: <u>ampere, A</u>.

2) In a <u>single</u>, closed <u>loop</u> (like the one on the right) the current
has the same value <u>everywhere</u> in the circuit (see p.48).

3) <u>Potential difference</u> (or voltage) is the <u>driving force</u> that <u>pushes</u>
the charge round. Unit of potential difference: <u>volt, V</u>.

4) <u>Resistance</u> is anything in the circuit which
<u>slows the flow down</u>. Unit of resistance: <u>ohm, Ω</u>.

5) The current flowing <u>through a component</u> depends on the <u>potential</u>
<u>difference</u> across it and the <u>resistance</u> of the component (see next page).

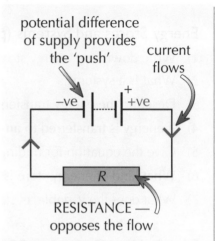

potential difference
of supply provides
the 'push'

current
flows

−ve　+ve

R

RESISTANCE —
opposes the flow

> The <u>greater the resistance</u> across a component, the <u>smaller the current</u> that
> flows through it (for a given potential difference across the component).

Total Charge Through a Circuit Depends on Current and Time

1) The <u>size</u> of the <u>current</u> is the <u>rate of flow</u> of <u>charge</u>. When <u>current</u> (*I*) flows past a point in a circuit
for a length of <u>time</u> (*t*) then the <u>charge</u> (*Q*) that has passed is given by this formula:

Charge (C) = Current (A) × Time (s)

$$Q = It$$

$$\frac{Q}{I \times t}$$

2) <u>Current</u> is measured in <u>amperes</u> (A),
<u>charge</u> is measured in <u>coulombs</u> (C),
<u>time</u> is measured in <u>seconds</u> (s).

3) <u>More charge</u> passes around the
circuit when a <u>bigger current</u> flows.

EXAMPLE:

**A battery charger passes a current of 2.0 A
through a cell over a period of 2.5 hours.
How much charge is transferred to the cell?**

$Q = It = 2.0 \times (2.5 \times 60 \times 60) = 18\ 000$ C

Learn these Circuit Diagram Symbols

You need to be able to <u>understand circuit diagrams</u> and draw them using the <u>correct symbols</u>. Make sure all
the <u>wires</u> in your circuit are <u>straight lines</u> and that the circuit is <u>closed</u>, i.e. you can follow a wire from one
end of the power supply, through any components, to the other end of the supply (ignoring any <u>switches</u>).

Cell	Battery	Switch open	Switch closed	Filament lamp (or bulb)	Fuse	LED
Resistor	Variable resistor	Ammeter	Voltmeter	Diode	LDR	Thermistor

Resistance

Prepare yourself to meet one of the most <u>important equations</u> in electronics. It's all about <u>resistance</u>, <u>current</u> and <u>potential difference</u>... Now if that doesn't tempt you on to read this page, I don't know what will.

There's a Formula Linking **Potential Difference** and **Resistance**

The formula linking <u>potential difference</u>, <u>current</u> and <u>resistance</u> is very useful (and pretty common):

You may see potential difference called voltage.

Potential difference (V)= Current (A) × Resistance (Ω)	**V = IR**

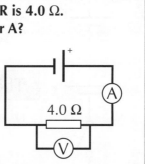

Use this formula triangle to rearrange the equation. Just cover up the thing you're trying to find, and what's left visible is the formula you're after.

EXAMPLE:

Voltmeter V reads 6.0 V and resistor R is 4.0 Ω. What is the current through ammeter A?

1) Use the formula triangle for $V = I \times R$.

2) We need to find I, so the version we need is $I = V \div R$.

3) The answer is then:

$$I = 6.0 \div 4.0 = 1.5 \text{ A}$$

4.0 Ω

Ohmic Conductors Have a **Constant Resistance**

For some components, as the <u>current</u> through them is changed, the <u>resistance</u> of the component changes as well.

1) The <u>resistance</u> of <u>ohmic conductors</u> (e.g. a <u>wire</u> or a <u>resistor</u>) doesn't change with the <u>current</u>. At a <u>constant temperature</u>, the current flowing through an ohmic conductor is <u>directly proportional</u> to the potential difference across it. (R is constant in $V = IR$.)

2) The resistance of some resistors and components does change, e.g. a <u>filament lamp</u> or a <u>diode</u>.

3) When an <u>electrical charge</u> flows through a filament lamp, it <u>transfers</u> some energy to the <u>thermal energy store</u> (p.17) of the filament, which is designed to <u>heat up</u>. Resistance increases with <u>temperature</u>, so as the <u>current</u> increases, the filament lamp heats up more and the resistance increases.

4) For <u>diodes</u>, the resistance depends on the <u>direction</u> of the current. They will happily let current flow in one direction, but have a <u>very high resistance</u> if it is <u>reversed</u>.

Resistance can be temperamental when it comes to temperature...

Remember that <u>ohmic conductors</u> will only have a <u>constant resistance</u> at a <u>constant temperature</u>. In general, <u>resistance increases with temperature</u> (though there are some exceptions, like thermistors — see p.45). So if the temperature is changing, the resistance of your component will be changing too.

Investigating Resistance

Resistance can depend on a number of factors. Here's an experiment you can do to investigate one of them — how the resistance varies with the length of the conductor.

You Can **Investigate** the Factors Affecting **Resistance**

The resistance of a circuit can depend on a number of factors, like whether components are in series or parallel, p.50, or the length of wire used in the circuit. You can investigate the effect of wire length using the circuit below.

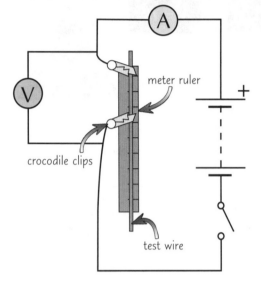

crocodile clips
meter ruler
test wire

The **Ammeter**

1) Measures the current (in amps) flowing through the test wire.

2) The ammeter must always be placed in series with whatever you're investigating.

The **Voltmeter**

See pages 47-49 for more on series and parallel circuits.

1) Measures the potential difference (or pd) across the test wire (in volts).

2) The voltmeter must always be placed in parallel around whatever you're investigating (p.49) — NOT around any other bit of the circuit, e.g. the battery.

Method

1) Attach a crocodile clip to the wire, level with 0 cm on the ruler.

2) Attach the second crocodile clip to the wire, e.g. 10 cm away from the first clip. Write down the length of the wire between the clips.

A thin wire will give you the best results. Make sure it's as straight as possible so your length measurements are accurate.

3) Close the switch, then record the current through the wire and the pd across it.

4) Open the switch, then move the second crocodile clip, e.g. another 10 cm, along the wire. Close the switch again, then record the new length, current and pd.

The wire may heat up during the experiment, which will affect its resistance (p.41). Leave the switch open for a bit between readings to let the circuit cool down.

5) Repeat this for a number of different lengths of the test wire.

6) Use your measurements of current and pd to calculate the resistance for each length of wire, using $R = V \div I$ (from $V = IR$).

7) Plot a graph of resistance against wire length and draw a line of best fit.

8) Your graph should be a straight line through the origin, meaning resistance is directly proportional to length — the longer the wire, the greater the resistance.

9) If your graph doesn't go through the origin, it could be because the first clip isn't attached exactly at 0 cm, so all of your length readings are a bit out. This is a systematic error (p.7).

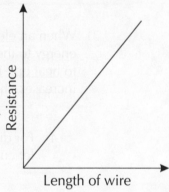

Resistance

Length of wire

PRACTICAL TIP

Be careful with the temperature of the wire...

If a large current flows through a wire, it can cause it to heat up (there's more on this on page 41). So use a low pd to stop it getting too hot and turn off the circuit between readings to let it cool.

I-V Characteristics

You've met <u>ohmic conductors</u> already on page 41, but most circuit components <u>aren't</u> ohmic.
You can see how circuit components behave by plotting an <u>*I-V* characteristic</u>.

Three Very Important *I-V* Characteristics

1) The term '<u>*I-V* characteristic</u>' refers to a <u>graph</u> which shows how the <u>current</u> (*I*) flowing through a component changes as the <u>potential difference</u> (*V*) across it is increased.

2) <u>Linear</u> components (e.g. an ohmic conductor) have an *I-V* characteristic that's a <u>straight line</u>.

3) <u>Non-linear</u> components (e.g. a filament lamp or a diode) have a <u>curved</u> *I-V* characteristic.

You can do this <u>experiment</u> to find a component's *I-V* characteristic:

Method

1) Set up the <u>test circuit</u> shown on the right.

2) Begin to vary the variable resistor. This alters the <u>current</u> flowing through the circuit and the <u>potential difference</u> across the <u>component</u>.

3) Take several <u>pairs of readings</u> from the <u>ammeter</u> and <u>voltmeter</u> to see how the <u>potential difference</u> across the component <u>varies</u> as the <u>current changes</u>. Repeat each reading twice more to get an <u>average</u> pd at each current.

4) <u>Swap</u> over the wires connected to the cell, so the <u>direction of the current</u> is reversed.

5) <u>Plot a graph</u> of <u>current against voltage</u> for the component.

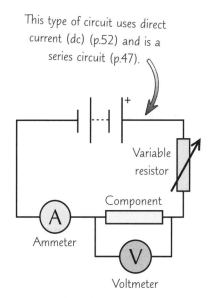

This type of circuit uses direct current (dc) (p.52) and is a series circuit (p.47).

Variable resistor

Component

A — Ammeter

V — Voltmeter

The *I-V* characteristics you get for an <u>ohmic conductor</u>, <u>filament lamp</u> and <u>diode</u> should look like this:

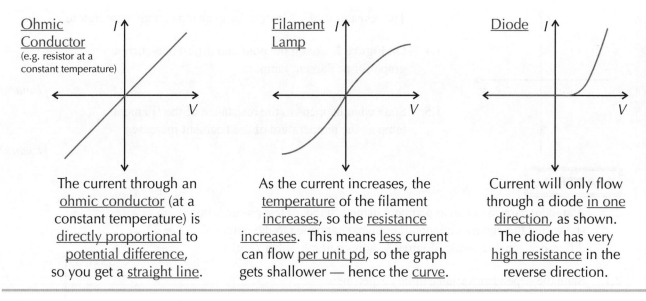

<u>Ohmic Conductor</u>
(e.g. resistor at a constant temperature)

The current through an <u>ohmic conductor</u> (at a constant temperature) is <u>directly proportional</u> to <u>potential difference</u>, so you get a <u>straight line</u>.

<u>Filament Lamp</u>

As the current increases, the <u>temperature</u> of the filament <u>increases</u>, so the <u>resistance increases</u>. This means <u>less</u> current can flow <u>per unit pd</u>, so the graph gets shallower — hence the <u>curve</u>.

<u>Diode</u>

Current will only flow through a diode <u>in one direction</u>, as shown. The diode has very <u>high resistance</u> in the reverse direction.

Since *V = IR*, you can calculate the <u>resistance</u> at any <u>point</u> on the *I-V* characteristic by calculating <u>*R = V ÷ I*</u>.

EXAM TIP

You may be asked to interpret an *I-V* characteristic...

Make sure you take care when reading values off the graph. Pay close attention to the <u>axes</u>, and make sure you've converted all values to the <u>correct units</u> before you do any calculations.

Warm-Up & Exam Questions

Phew — circuits aren't the easiest thing in the world, are they? Make sure you've understood the last few pages by trying these questions. If you get stuck, just go back and re-read the relevant page.

Warm-Up Questions

1) What are the units of resistance?
2) How does current through a component vary with resistance for a fixed potential difference?
3) Draw the symbol for a light-emitting diode (LED).
4) Give an example of an ohmic conductor.
5) How should a voltmeter be connected in a circuit to measure the pd across a component?
6) What is an *I-V* characteristic?

Exam Questions

1 **Figure 1** shows is a circuit diagram for a standard test circuit. When the switch is closed, the ammeter reads 0.30 A and the voltmeter reads 1.5 V.

Grade 6-7

Figure 1

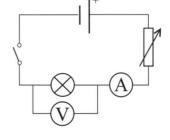

1.1 Calculate the resistance of the filament lamp.

[3 marks]

1.2 The switch is closed for 35 seconds.
Calculate the total charge that flows through the filament lamp.

[2 marks]

1.3 The variable resistor is used to increase the resistance in the circuit.
Describe how this will affect the current flowing through the circuit.

[1 mark]

Figure 2

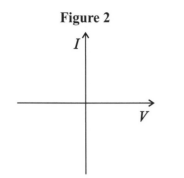

The resistance of a filament lamp changes with temperature.

1.4 On **Figure 2**, sketch the potential difference-current graph for a filament lamp.

[1 mark]

1.5 State what happens to the resistance of the filament lamp as the temperature of the filament increases.

[1 mark]

PRACTICAL

2 A student carried out an experiment using a standard test circuit where she varied the current and monitored what happened to the potential difference across a diode. **Figure 3** shows a graph of her results.

Grade 6-7

Figure 3

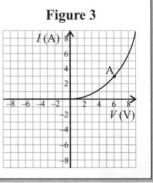

2.1 State the dependent variable in this experiment.

[1 mark]

2.2 Explain why the graph in **Figure 3** shows zero current for negative pds.

[1 mark]

2.3 Calculate the resistance of the diode at the point marked A.

[4 marks]

Circuit Devices

You might consider yourself a bit of an expert in <u>circuit components</u> — you're enlightened about bulbs, you're switched on to switches... Just make sure you know these ones as well — they're a bit trickier.

A **Light-Dependent Resistor** or **"LDR"**

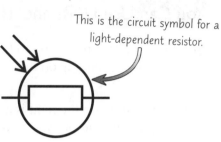

This is the circuit symbol for a light-dependent resistor.

1) An LDR is a resistor that is <u>dependent</u> on the <u>intensity</u> of <u>light</u>. In <u>bright light</u>, the resistance <u>falls</u>.

2) In <u>darkness</u>, the resistance is <u>highest</u>.

3) They have lots of applications including <u>automatic night lights</u>, outdoor lighting and <u>burglar detectors</u>.

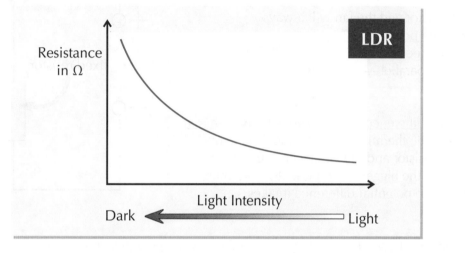

Thermistor Resistance Decreases as **Temperature Increases**

1) A <u>thermistor</u> is a <u>temperature dependent</u> resistor.

2) In <u>hot</u> conditions, the resistance <u>drops</u>.

3) In <u>cool</u> conditions, the resistance goes <u>up</u>.

4) Thermistors make useful <u>temperature detectors</u>, e.g. <u>car engine</u> temperature sensors and electronic <u>thermostats</u>.

This is the circuit symbol for a thermistor.

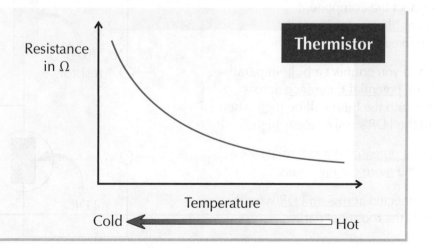

Thermistors and LDRs have many applications...

And they're not just limited to the examples on this page. Oh no. For example, LDRs are used in <u>digital cameras</u> to control how long the <u>shutter</u> should stay open for. If the <u>light level</u> is <u>low</u>, changes in the <u>resistance</u> cause the shutter to <u>stay open for longer</u> than if the light level was higher. How interesting.

Sensing Circuits

Now you've learnt about what <u>LDRs</u> and <u>thermistors</u> do, it's time to take a look at how they're put to use.

You Can Use LDRs and Thermistors in **Sensing Circuits**

<u>Sensing circuits</u> can be used to turn on or <u>increase the power</u> to components depending on the conditions that they are in.

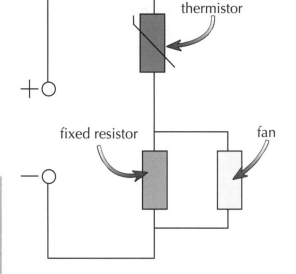

1) The circuit on the right is a <u>sensing circuit</u> used to operate a fan in a room.

2) The fixed resistor and the fan will always have the <u>same potential difference</u> across them (because they're connected in parallel — see p.49).

3) The <u>potential difference</u> of the power supply is <u>shared out</u> between the thermistor and the loop made up of the fixed resistor and the fan according to their <u>resistances</u> — the <u>bigger</u> a component's resistance, the <u>more</u> of the potential difference it takes.

4) As the room gets <u>hotter</u>, the resistance of the thermistor <u>decreases</u> and it takes a <u>smaller share</u> of the potential difference from the power supply. So the potential difference across the fixed resistor and the fan <u>rises</u>, making the fan go faster.

You can Connect the **Component Across** the **Variable Resistor**

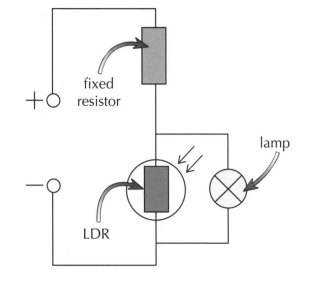

1) You can connect the component <u>across</u> the <u>variable resistor</u> instead of across the fixed resistor.

2) For example, if you connect a <u>bulb</u> in parallel to an <u>LDR</u>, the <u>potential difference</u> across both the LDR and the bulb will be <u>high</u> when it's <u>dark</u> and the LDR's resistance is <u>high</u>.

3) The <u>greater the potential difference</u> across a component, the <u>more energy</u> it gets.

4) So a <u>bulb</u> connected <u>across an LDR</u> would get <u>brighter</u> as the room got <u>darker</u>.

Sensing circuits react to changes in the surroundings...

<u>Sensing circuits</u> are a useful <u>application</u> of <u>thermistors</u> and <u>LDRs</u>, but they can be tricky to make sense of. They rely on the properties of <u>series</u> and <u>parallel circuits</u> — <u>read on</u> to learn all about them.

Series Circuits

You need to be able to tell the if components are connected in series or parallel <u>just by looking at circuit diagrams</u>. You also need to know the <u>rules</u> about what happens with both types. Read on to find out more.

Series Circuits — **All** or **Nothing**

1) In <u>series circuits</u>, the different components are connected <u>in a line</u>, <u>end to end</u>, between the +ve and –ve of the power supply (except for <u>voltmeters</u>, which are always connected <u>in parallel</u>, but they don't count as part of the circuit).

2) If you remove or disconnect <u>one</u> component, the circuit is <u>broken</u> and they all <u>stop</u>. This is generally <u>not very handy</u>, and in practice <u>very few things</u> are connected in series.

3) You can use the following rules to <u>design</u> series circuits to <u>measure quantities</u> and test components (e.g. the <u>test circuits</u> on p.43 and p.50 and the <u>sensing circuits</u> on the last page).

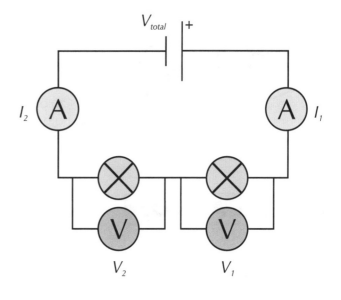

Cell Potential Differences **Add Up**

1) There is a bigger potential difference when more cells are in series, provided the cells are all <u>connected</u> the <u>same way</u>.

2) For example, when two batteries of voltage 1.5 V are <u>connected in series</u> they supply a <u>total</u> of 3 V.

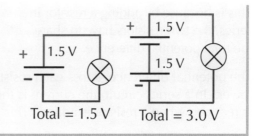

Total Potential Difference is **Shared**

In series circuits, the <u>total potential difference</u> of the <u>supply</u> is <u>shared</u> between the various <u>components</u>. So the <u>potential differences</u> round a series circuit <u>always add up</u> to the <u>source potential difference</u>:

$$V_{total} = V_1 + V_2 + ...$$

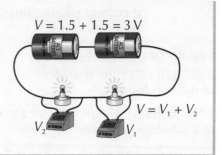

There are two main types of circuit — series and parallel...

Remember, <u>ammeters</u> should always be <u>connected in series</u>, and <u>voltmeters</u> should always be <u>connected in parallel</u>. These components don't count towards how you define a circuit — you can have a <u>parallel circuit</u> (p.49) with <u>ammeters</u> connected in <u>series</u>, or a <u>series circuit</u> with <u>voltmeters</u> connected <u>across</u> components.

Series Circuits

We're not done with <u>series circuits</u> yet. Here's the low-down on <u>current</u> and <u>resistance</u>...

Current is the **Same Everywhere**

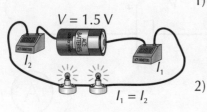

$V = 1.5\,V$

I_2

I_1

$I_1 = I_2$

1) In series circuits the <u>same current</u> flows through <u>all components</u>, i.e:

$$I_1 = I_2 = \,...$$

2) The <u>size</u> of the current is determined by the <u>total potential difference</u> of the cells and the <u>total resistance</u> of the circuit: i.e. $I = V \div R$.

Resistance **Adds Up**

1) In series circuits the <u>total resistance</u> of two components is just the <u>sum</u> of their resistances:

$$R_{total} = R_1 + R_2$$

2) This is because by <u>adding a resistor</u> in series, the two resistors have to <u>share</u> the total potential difference.

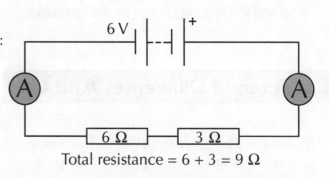

6 V

6 Ω   3 Ω

Total resistance = 6 + 3 = 9 Ω

3) The potential difference across each resistor is <u>lower</u>, so the <u>current</u> through each resistor is also lower. In a series circuit, the current is the <u>same everywhere</u> so the total current in the circuit is <u>reduced</u> when a resistor is added. This means the total <u>resistance</u> of the circuit <u>increases</u>.

4) The <u>bigger</u> a component's <u>resistance</u>, the bigger its <u>share</u> of the <u>total potential difference</u>.

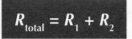 EXAMPLE:

For the circuit diagram on the right, calculate the current passing through the circuit.

1) First find the <u>total resistance</u> by <u>adding together</u> the resistance of the two resistors.

2) Then <u>rearrange</u> $V = IR$ and <u>substitute</u> in the values you have.

$R_{total} = 2 + 3 = 5\,\Omega$

$I = V \div R$
$= 20 \div 5$
$= 4\,A$

2 Ω   3 Ω

20 V

Series circuits aren't used very much in the real world...

Since series circuits put <u>all</u> components on the <u>same loop of wire</u>, and the current is the same through each component, if one <u>component breaks</u>, it'll <u>break the circuit</u>, and all other components will <u>stop working</u> too. Parallel circuits are much more useful and can avoid this problem — as you're about to find out...

Parallel Circuits

Parallel circuits can be a little bit trickier to wrap your head around, but they're much more useful than series circuits. Most electronics use a combination of series and parallel circuitry.

Parallel Circuits — Independence and Isolation

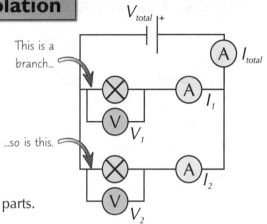

1) In parallel circuits, each component is separately connected to the +ve and –ve of the supply (except ammeters, which are always connected in series).

2) If you remove or disconnect one of them, it will hardly affect the others at all.

3) This is obviously how most things must be connected, for example in cars and in household electrics. You have to be able to switch everything on and off separately.

4) Everyday circuits often include a mixture of series and parallel parts.

Potential Difference is the **Same** Across All **Components**

1) In parallel circuits all components get the full source pd, so the voltage is the same across all components:

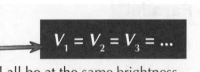

$$V_1 = V_2 = V_3 = ...$$

2) This means that identical bulbs connected in parallel will all be at the same brightness.

Current is **Shared** Between Branches

1) In parallel circuits the total current flowing around the circuit is equal to the total of all the currents through the separate components:

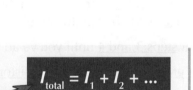

$$I_{total} = I_1 + I_2 + ...$$

2) In a parallel circuit, there are junctions where the current either splits or rejoins. The total current going into a junction has to equal the total current leaving it.

3) If two identical components are connected in parallel then the same current will flow through each component.

Adding a Resistor in Parallel **Reduces** the **Total Resistance**

1) If you have two resistors in parallel, their total resistance is less than the resistance of the smallest of the two resistors.

2) This can be tough to get your head around, but think about it like this:

- In parallel, both resistors have the same potential difference across them as the source.
- This means the 'pushing force' making the current flow is the same as the source potential difference for each resistor that you add.
- But by adding another loop, the current has more than one direction to go in.
- This increases the total current that can flow around the circuit. Using $V = IR$, an increase in current means a decrease in the total resistance of the circuit.

Circuits and Resistance

You saw on page 42 how the length of the wire used in a circuit affects its resistance. Now it's time to do an experiment to see how placing resistors in series or in parallel can affect the resistance of the whole circuit.

You Can **Investigate** Adding **Resistors** in **Series**...

1) First, you'll need to find at least four identical resistors.

2) Then build the circuit shown on the right using one of the resistors. Make a note of the potential difference of the battery (V).

3) Measure the current through the circuit using the ammeter. Use this to calculate the resistance of the circuit using $R = V \div I$.

4) Add another resistor, in series with the first.

5) Again, measure the current through the circuit and use this and the potential difference of the battery to calculate the overall resistance of the circuit.

6) Repeat steps 4 and 5 until you've added all of your resistors.

7) Plot a graph of the number of resistors against the total resistance of the circuit (see below).

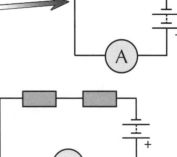

... or in **Parallel**

1) Using the same equipment as before (so the experiment is a fair test), build the same initial circuit.

2) Measure the total current through the circuit and calculate the resistance of the circuit using $R = V \div I$ (again, V is the potential difference of the battery).

3) Next, add another resistor, in parallel with the first.

4) Measure the total current through the circuit and use this and the potential difference of the battery to calculate the overall resistance of the circuit.

5) Repeat steps 3 and 4 until you've added all of your resistors.

6) Plot a graph of the number of resistors in the circuit against the total resistance.

Your Results Should **Match** the **Resistance Rules**

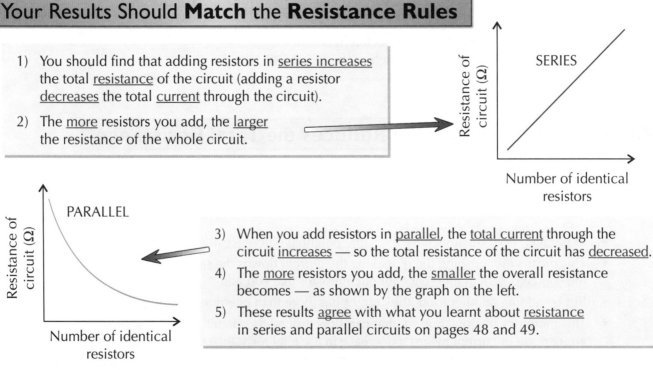

1) You should find that adding resistors in series increases the total resistance of the circuit (adding a resistor decreases the total current through the circuit).

2) The more resistors you add, the larger the resistance of the whole circuit.

3) When you add resistors in parallel, the total current through the circuit increases — so the total resistance of the circuit has decreased.

4) The more resistors you add, the smaller the overall resistance becomes — as shown by the graph on the left.

5) These results agree with what you learnt about resistance in series and parallel circuits on pages 48 and 49.

Warm-Up & Exam Questions

Time to check and see what you can remember about those circuit devices, parallel and series circuits.

Warm-Up Questions

1) Give one use of a light-dependent resistor (LDR).
2) What happens to the resistance of a thermistor as its temperature increases?
3) Draw a circuit diagram of a sensing circuit where a bulb gets brighter with decreasing temperature.
4) Give one practical disadvantage of series circuits.
5) How do you work out the total resistance in a series circuit?
6) Which has the higher total resistance: two resistors in series, or the same two resistors in parallel?
7) Sketch a graph to show how the number of identical resistors connected together in parallel affects the total resistance in a circuit.

Exam Questions

1 **Figure 1** shows a series circuit. (Grade 4-6)

Figure 1

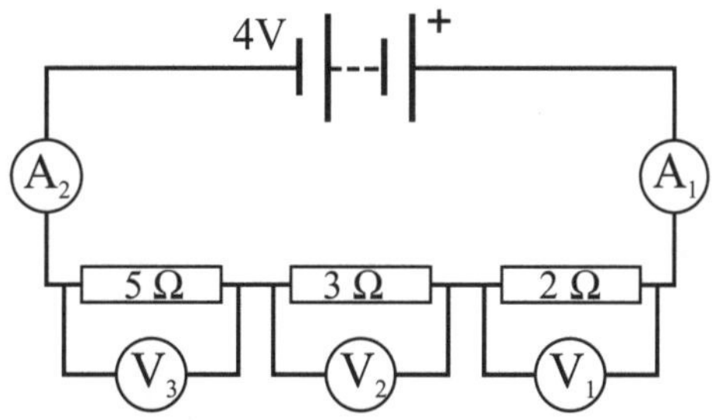

1.1 Calculate the total resistance in the circuit.

[2 marks]

1.2 The current through A_1 is 0.4 A.
What is the current through A_2? Explain your answer.

[2 marks]

1.3 V_1 reads 0.8 V and V_2 reads 1.2 V.
Calculate the reading on V_3.

[2 marks]

2 A parallel circuit is connected as shown in **Figure 2**. (Grade 7-9)

Figure 2

2.1 Give the reading on voltmeter V_1.

[1 mark]

2.2 Calculate the reading on ammeter A_1.

[3 marks]

2.3 Calculate the reading on ammeter A_2.

[2 marks]

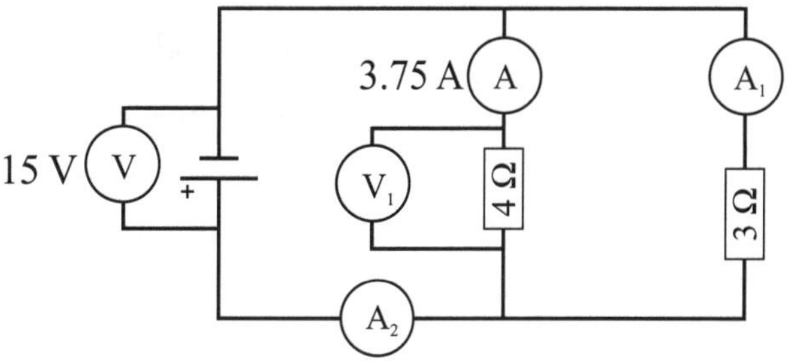

Electricity in the Home

Now you've learnt the basics of <u>electrical circuits</u>, it's time to see how <u>electricity</u> is used in <u>everyday life</u>.

Mains Supply is **ac**, **Battery** Supply is **dc**

1) There are two types of electricity supplies — <u>alternating current</u> (ac) and <u>direct current</u> (dc).

2) In <u>ac supplies</u> the current is <u>constantly</u> changing direction. <u>Alternating currents</u> are produced by <u>alternating potential difference</u> in which the <u>positive</u> and <u>negative</u> ends keep <u>alternating</u>.

3) The <u>UK mains supply</u> (the electricity in your home) is an ac supply at around <u>230 V</u>.

4) The frequency of the ac mains supply is <u>50 cycles per second</u> or <u>50 Hz</u> (hertz).

5) By contrast, cells and batteries supply <u>direct current</u> (dc).

6) <u>Direct current</u> is a current that is always flowing in the <u>same direction</u>. It's created by a <u>direct potential difference</u>.

Most Cables Have **Three** Separate **Wires**

1) Most electrical appliances are connected to the mains supply by <u>three-core</u> cables. This means that they have <u>three wires</u> inside them, each with a <u>core of copper</u> and a <u>coloured plastic coating</u>.

2) The <u>colour</u> of the insulation on each cable shows its <u>purpose</u>.

3) The colours are <u>always</u> the <u>same</u> for <u>every</u> appliance. This is so that it is easy to tell the different wires <u>apart</u>.

4) You need to know the <u>colour</u> of each wire, what each of them is <u>for</u> and what their <u>pd</u> is.

NEUTRAL WIRE — <u>blue</u>. The neutral wire <u>completes</u> the circuit — when the appliance is operating normally, current flows through the <u>live</u> and <u>neutral</u> wires. It is around <u>0 V</u>.

LIVE WIRE — <u>brown</u>. The live wire provides the <u>alternating potential difference</u> (at about <u>230 V</u>) from the mains supply.

EARTH WIRE — <u>green</u> and yellow. The earth wire is for protecting the wiring, and for safety — it stops the appliance casing from <u>becoming live</u>. It doesn't usually carry a current — only when there's a <u>fault</u>. It's <u>also</u> at 0 V.

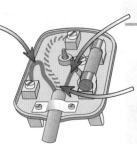

Touching the **Live Wire** Gives You an **Electric Shock**

1) Your <u>body</u> (just like the earth) is at <u>0 V</u>.

2) This means that if you touch the <u>live wire</u>, a <u>large potential difference</u> is produced across your body and a <u>current</u> flows through you.

3) This causes a large <u>electric shock</u> which could injure or even kill you.

4) Even if a plug socket or a light switch is turned <u>off</u> (i.e. the switch is <u>open</u>) there is still a <u>danger</u> of an electric shock. A current <u>isn't flowing</u>, but there is still a pd in the live wire. If you made <u>contact</u> with the live wire, your body would provide a <u>link</u> between the supply and the earth, so a <u>current</u> would flow <u>through you</u>.

5) <u>Any</u> connection between <u>live</u> and <u>earth</u> can be <u>dangerous</u>. If the link creates a <u>low resistance</u> path to earth, a huge current will flow, which could result in a fire.

Power of Electrical Appliances

You can think about <u>electrical circuits</u> in terms of <u>energy transfer</u> — the charge carriers take energy around the circuit. When they go through an electrical component energy is transferred to make the component work.

Energy is **Transferred** from Cells and Other **Sources**

1) You know from page 17 that a moving charge <u>transfers energy</u>. This is because the charge <u>does work against</u> the <u>resistance</u> of the circuit. (Work done is the <u>same</u> as energy transferred, p.18.)

2) Electrical appliances are designed to <u>transfer energy</u> to components in the circuit when a <u>current</u> flows.

Kettles transfer energy <u>electrically</u> from the mains ac supply to the <u>thermal</u> energy store of the heating element inside the kettle.	Energy is transferred <u>electrically</u> from the <u>battery</u> of a handheld fan to the <u>kinetic</u> energy store of the fan's motor.

3) Of course, <u>no</u> appliance transfers <u>all</u> energy completely usefully. The <u>higher</u> the <u>current</u>, the more energy is transferred to the <u>thermal</u> energy stores of the components (and then the surroundings). You can calculate the <u>efficiency</u> of any electrical appliance — see p.28.

Energy Transferred Depends on the **Power**

1) The <u>total</u> energy transferred by an appliance depends on <u>how long</u> the appliance is on for and its <u>power</u>.

2) The <u>power</u> of an appliance is the energy that it <u>transfers per second</u>. So the <u>more</u> energy it transfers in a given time, the <u>higher</u> its power.

3) The amount of <u>energy transferred by electrical work</u> is given by:

> **Energy transferred (J) = Power (W) × Time (s)** $E = Pt$

This equation should be familiar from page 23.

EXAMPLE:

A 600 W microwave is used for 5 minutes. How long (in minutes) would a 750 W microwave take to do the same amount of work?

1) Calculate the <u>energy transferred</u> by the <u>600 W</u> microwave in <u>five minutes</u>.
$E = Pt = 600 × (5 × 60)$
$= 180\ 000\ J$

Remember that the time must be in seconds.

2) <u>Rearrange</u> $E = Pt$ and <u>sub in</u> the <u>energy</u> you calculated and the <u>power</u> of the 750 W microwave.
$t = E ÷ P$
$= 180\ 000 ÷ 750 = 240\ s$

3) <u>Convert</u> the time back to <u>minutes</u>.
$240 ÷ 60 = 4$ minutes

So the 750 W microwave would take 4 minutes to do the same amount of work.

4) Appliances are often given a <u>power rating</u> — they're labelled with the <u>maximum</u> safe power that they can operate at. You can usually take this to be their <u>maximum operating power</u>.

5) The power rating tells you the <u>maximum</u> amount of <u>energy</u> transferred between stores <u>per second</u> when the appliance is in use.

6) This helps customers choose between models — the <u>lower</u> the power rating, the <u>less</u> electricity an appliance uses in a given time and so the <u>cheaper</u> it is to run.

7) But a higher power <u>doesn't</u> necessarily mean that it transfers <u>more</u> energy <u>usefully</u>. An appliance may be <u>more powerful</u> than another, <u>but less efficient</u>, meaning that it might still only transfer the <u>same amount</u> of energy (or even <u>less</u>) to useful stores (see p.28).

More on Power

As you've seen, the power of a device tells you how much energy it transfers per second.
In electrical systems, there are a load of useful formulas you can use to calculate energy and power.

Potential Difference is **Energy Transferred** per **Charge Passed**

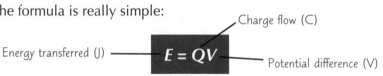

1) When an electrical charge goes through a change
   in potential difference, energy is transferred.

2) Energy is supplied to the charge at the power source
   to 'raise' it through a potential.

3) The charge gives up this energy when it 'falls' through any
   potential drop in components elsewhere in the circuit.

4) The formula is really simple:

   Charge flow (C)

   Energy transferred (J) —— $E = QV$ —— Potential difference (V)

   Charges gaining energy at the battery. Charges releasing energy in resistors.

5) That means that a battery with a bigger pd will supply more energy to the circuit for every
   coulomb of charge which flows round it, because the charge is raised up "higher" at the start.

EXAMPLE:

The motor in an electric toothbrush is attached to a 3 V battery.
140 C of charge passes through the circuit as it is used.
Calculate the energy transferred.

$E = QV = 140 \times 3 = 420$ J

This energy is transferred to the kinetic energy store of the motor, as well as to the thermal energy stores of the surroundings.

Power Also Depends on **Current** and **Potential Difference**

1) As well as energy transferred in a given time, the power of an appliance can be found with:

Power (W) = Potential difference (V) × Current (A)	$P = VI$

EXAMPLE:

A 1.0 kW hair dryer is connected to a 230 V supply. Calculate the
current through the hair dryer. Give your answer to two significant figures.

1) Rearrange the equation for current. $I = P \div V$
2) Make sure your units are correct. 1.0 kW = 1000 W
3) Then just stick in the numbers that you have. $I = 1000 \div 230 = 4.34... = 4.3$ A (to 2 s.f.)

2) You can also find the power if you don't know the potential difference.
   To do this, stick $V = IR$ from page 41 into $P = VI$, which gives you: Resistance (Ω) $P = I^2R$

Power is measured in watts, W — one W is equal to one J/s...

Remember, the power rating of an electrical appliance is the amount of energy transferred to the appliance
per second, not the amount that it transfers to useful energy stores. Two appliances with the same power
rating won't necessarily work as well as each other — it'll depend on their efficiencies (see page 28).

The National Grid

The national grid is a giant web of wires that covers the whole of Britain, getting electricity from power stations to homes everywhere. Whoever you pay for your electricity, it's the national grid that gets it to you.

Electricity is Distributed via the National Grid

1) The national grid is a giant system of cables and transformers (p.171) that covers the UK and connects power stations to consumers (anyone who is using electricity).

2) The national grid transfers electrical power from power stations anywhere on the grid (the supply) to anywhere else on the grid where it's needed (the demand) — e.g. homes and industry.

Electricity Production has to Meet Demand

1) Throughout the day, electricity usage (the demand) changes. Power stations have to produce enough electricity for everyone to have it when they need it.

2) They can predict when the most electricity will be used though. Demand increases when people get up in the morning, come home from school or work and when it starts to get dark or cold outside. Popular events like a sporting final being shown on TV could also cause a peak in demand.

3) Power stations often run at well below their maximum power output, so there's spare capacity to cope with a high demand, even if there's an unexpected shut-down of another station.

4) Lots of smaller power stations that can start up quickly are also kept in standby just in case.

Energy demands are ever increasing...

The national grid has been working since the 1930s and has gone through many changes and updates since then to meet increasing energy demands. Using energy-efficient appliances and switching unneeded lights off are some ways we might ensure that supply and demand stay in balance. It'll do wonders for your electricity bills too, as I'm sure your parents often remind you.

The National Grid

To transfer electricity <u>efficiently</u>, the national grid makes use of some clever tech called <u>transformers</u>.

The National Grid Uses a **High pd** and a **Low Current**

1) To transmit the <u>huge</u> amount of <u>power</u> needed, you need either a <u>high potential difference</u> or a <u>high current</u> (as *P = VI*, page 54).

2) The <u>problem</u> with a <u>high current</u> is that you lose <u>loads of energy</u> as the wires <u>heat up</u> and energy is transferred to the <u>thermal</u> energy store of the <u>surroundings</u>.

3) It's much <u>cheaper</u> to <u>boost the pd</u> up <u>really high</u> (to 400 000 V) and keep the current <u>relatively low</u>.

4) For a given <u>power</u>, increasing the pd <u>decreases</u> the <u>current</u>, which decreases the <u>energy lost</u> by heating the wires and the surroundings. This makes the national grid an <u>efficient</u> way of transferring energy.

Remember that power is the energy transferred in a given time, so a higher power means more energy transferred.

Potential Difference is Changed by a Transformer

1) To get the potential difference to 400 000 V to transmit power requires <u>transformers</u> as well as <u>big pylons</u> with <u>huge insulators</u> — but it's <u>still cheaper</u>.

2) The transformers have to <u>step</u> the potential difference <u>up</u> at one end, for <u>efficient transmission</u>, and then bring it back down to <u>safe, usable levels</u> at the other end.

For more on how transformers work, have a look at page 171.

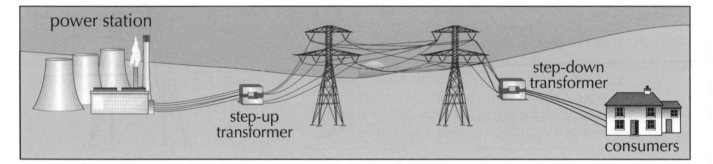

3) The <u>potential difference</u> is <u>increased</u> ('<u>stepped up</u>') using a <u>step-up transformer</u>.

4) It's then <u>reduced</u> again ('<u>stepped down</u>') for domestic use using a <u>step-down transformer</u>.

5) Transformers are <u>almost 100% efficient</u>. So you can assume that the <u>input power</u> is <u>equal</u> to the <u>output power</u>. Using *P = VI* from page 54, you can write this as:

V_s = pd across secondary coil
I_s = current through secondary coil

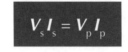
$$V_s I_s = V_p I_p$$

V_p = pd across primary coil
I_p = current through primary coil

$V_s \times I_s$ is the power output at the secondary coil. $V_p \times I_p$ is the power input at the primary coil.

The national grid — it's a powerful thing...

The key to the <u>efficiency</u> of the <u>national grid</u> is the power equation, <u>*P = VI*</u> (see page 54). Since <u>power</u> is <u>proportional</u> to both <u>potential difference</u> and <u>current</u>, if you have a <u>constant</u> power, but <u>increase</u> the potential difference using a transformer, the current must <u>decrease</u>. And vice versa.

Warm-Up & Exam Questions

Who knew there was so much to learn about electricity in the home and across the country?
See if it's switched on a lightbulb in your brain by trying out these questions.

Warm-Up Questions

1) What are the three core cables that connect electrical appliances to the mains supply?
2) What is the main energy transfer when electric current flows through an electric kettle?
3) What is the equation linking power, current and resistance?
4) What is the national grid?

Exam Questions

1   An electrical cable has become frayed so that the metal part of the live wire is exposed.   *(Grade 4-6)*
Explain why you would get an electric shock if you touched it.

[3 marks]

2   **Table 1** shows the power and potential difference ratings for two kettles.   *(Grade 6-7)*

Table 1

	Power (kW)	**Potential Difference (V)**
Kettle A	2.8	230
Kettle B	3.0	230

2.1   State the equation linking power, potential difference and current.

[1 mark]

2.2   Calculate the current drawn from the mains supply by kettle A. State the correct unit.

[4 marks]

2.3   A student is deciding whether to buy kettle A or kettle B.
She wants to buy the kettle that boils water faster. Both kettles have an efficiency of 90%.
Suggest which kettle she should choose. Explain your answer.

[2 marks]

3   The national grid transmits electricity from power stations   *(Grade 6-7)*
to homes and businesses all over the country.

3.1   Explain why the national grid uses step-up transformers.

[3 marks]

3.2   Electricity is passed through a step-down transformer. The potential difference across the primary coil is 4.00×10^5 V and the current through the primary coil is 2.00×10^3 A. The potential difference across the secondary coil is 2.75×10^5 V. Calculate the current in the secondary coil.

[3 marks]

4   A current of 0.5 A passes through a torch bulb. The torch is powered by a 3.0 V battery.   *(Grade 7-9)*

4.1   The torch is on for half an hour.
Calculate the amount of energy transferred from the battery in this time.

[4 marks]

4.2   Calculate how much charge passes through the torch in half an hour.

[3 marks]

Static Electricity

Static electricity is all about charges which are <u>not</u> free to move, e.g. in insulating materials.

Build-up of **Static** is Caused by **Friction**

1) When certain <u>insulating</u> materials are <u>rubbed</u> together, negatively charged electrons will be <u>scraped off one</u> and <u>dumped</u> on the other.

2) This will leave the materials <u>electrically charged</u>, with a <u>positive</u> static charge on one and an <u>equal negative</u> static charge on the other.

3) <u>Which way</u> the electrons are transferred <u>depends</u> on the <u>two materials</u> involved.

4) The classic examples are <u>polythene</u> and <u>acetate</u> rods being rubbed with a <u>cloth duster</u> (shown below).

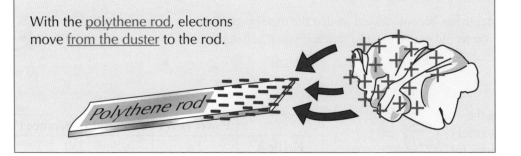

With the <u>polythene rod</u>, electrons move <u>from the duster</u> to the rod.

Polythene rod

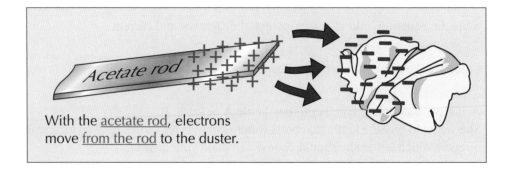

Acetate rod

With the <u>acetate rod</u>, electrons move <u>from the rod</u> to the duster.

Only Electrons Move — Never Positive Charges

1) <u>Watch out for this in exams</u>. Both positive (+ve) and negative (–ve) electrostatic charges are only ever produced by the movement of <u>electrons</u>. The positive charges <u>definitely do not move</u>!

2) A positive static charge is always caused by electrons <u>moving</u> away elsewhere. The material that <u>loses</u> the electrons loses some negative charge, and is <u>left with an equal positive charge</u>. Don't forget.

It shouldn't shock you to learn that static charges are everywhere...

You'll probably be pretty familiar with <u>static charges</u> from your day to day life. Whenever taking off a jumper or brushing your hair makes your hair stand on end — that's <u>static electricity</u>. It's also the reason you can get zapped by the car door after a long journey. The 'zap' is a spark — read on to find out more.

Static Electricity

When lots of static charge builds up in one place, it often ends with a <u>spark</u> or a <u>shock</u>, as they finally move.

Too Much Static Causes **Sparks**

1) As <u>electric charge</u> builds on an object, the <u>potential difference</u> between the object and the earth (which is at <u>0 V</u>) increases.

For more on how sparks actually jump across gaps, see the next page.

2) If the potential difference gets <u>large enough</u>, electrons can <u>jump</u> across the <u>gap</u> between the charged object and the earth — this is the <u>spark</u>.

3) They can also <u>jump</u> to any <u>earthed conductor</u> that is nearby — which is why <u>you</u> can get <u>static shocks</u> getting out of a car. A charge builds up on the car's <u>metal frame</u>, and when you touch the car, the <u>charge</u> travels <u>through you</u> to earth.

4) This <u>usually</u> happens when the gap is fairly <u>small</u>. (But not always — <u>lightning</u> is just a really big spark.)

Like Charges Repel, **Opposite** Charges Attract

1) This is <u>easy</u> and, I'd have thought, <u>kind of obvious</u>. When two electrically charged objects are brought close together they <u>exert a force</u> on one another.

2) Two things with <u>opposite</u> electric charges are <u>attracted</u> to each other, while two things with the <u>same</u> electric charge will <u>repel</u> each other.

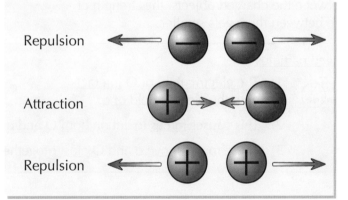

3) These forces get <u>weaker</u> the <u>further apart</u> the two things are.

4) These forces will cause the objects to <u>move</u> if they are able to do so. This is known as <u>electrostatic attraction / repulsion</u> and is a <u>non-contact</u> force (the objects don't need to touch, p.87).

5) One way to see this force is to <u>suspend</u> a <u>rod</u> with a <u>known charge</u> from a piece of string (so it is free to <u>move</u>). Placing an object with the <u>same charge</u> nearby will <u>repel</u> the rod — the rod will <u>move away</u> from the object. An <u>oppositely charged</u> object will cause the rod to move <u>towards</u> the object.

Electrostatic forces — like charges repel but opposites attract...

You could be asked to apply the idea of <u>electrostatic attraction</u> and <u>repulsion</u> in any situation in the exam. So remember, when an object becomes <u>charged</u> by being rubbed against another object, the objects will have <u>equal</u> and <u>opposite charges</u> to each other.

Electric Fields

As you've seen on the last page, electrical charges exert a force on each other. You can explain these forces in terms of electric fields. They can be used to explain things like sparks as well.

Electric **Charges** Create an **Electric Field**

1) An electric field is created around any electrically charged object.

2) The closer to the object you get, the stronger the field is. (And the further you are from it, the weaker it is.)

Isolated means it's not interacting with anything.

3) You can show an electric field around an object using field lines. For example you can draw the field lines for an isolated, charged sphere:

- Electric field lines go from positive to negative.
- They're always at a right angle to the surface.
- The closer together the lines are, the stronger the field is — you can see that the further from a charge you go, the further apart the lines are and so the weaker the field is.

Charged Objects in an Electric Field feel a **Force**

1) When a charged object is placed in the electric field of another object, it feels a force.

2) This force causes the attraction or repulsion you saw on the previous page.

3) The force is caused by the electric fields of each charged object interacting with each other.

4) The force on an object is linked to the strength of the electric field it is in.

5) As you increase the distance between the charged objects, the strength of the field decreases and the force between them gets smaller.

Two oppositely charged particles

Force   Force

1) The electric field of Q interacts with the electric field of q.

2) This causes forces to act on both Q and q.

3) These forces move q and Q closer together.

Sparking Can Be **Explained** By **Electric Fields**

1) Sparks are caused when there is a high enough potential difference between a charged object and the earth (or an earthed object).

2) A high potential difference causes a strong electric field between the charged object and the earthed object.

3) The strong electric field causes electrons in the air particles to be removed (known as ionisation).

4) Air is normally an insulator, but when it is ionised it is much more conductive, so a current can flow through it. This is the spark.

Warm-Up & Exam Questions

Have you been paying close attention to the last three pages? Only one way to tell really — have a go at these delightful warm-up and exam questions. If you get any wrong, go back and read it all again.

Warm-Up Questions

1) What happens to an acetate rod when it is rubbed by a cloth?
2) Two objects are rubbed together and become charged.
   What particles move to cause the objects to become charged?
3) Do opposite charges attract or repel each other?
4) What type of objects have electric fields?
5) What happens to the strength of an electric field as you move closer to a charged object?

Exam Questions

1   A student rubbed a plastic sphere with a cloth, and the sphere became negatively charged. **(Grade 4-6)**
     The charge on the sphere is –1.6 nC.

1.1 State the charge on the cloth.

[1 mark]

1.2 Describe the movement of charge which caused the sphere to become charged.

[1 mark]

An image of the charged sphere is shown in **Figure 1**.

Figure 1

A   B $\ominus$ C

1.3 Give the position (A, B or C) where a test charge would feel the greatest force.

[1 mark]

1.4 Draw an arrow on **Figure 1** showing the force a negative test charge would feel at that point.

[1 mark]

1.5 Describe how the presence of a strong electric field can cause a spark.

[2 marks]

2   Jane hangs an uncharged balloon from a thread. She brings a negatively charged
     polythene rod towards the balloon. **Figure 2** shows how the positive and   **(Grade 7-9)**
     negative charges in the balloon rearrange themselves when she does this.

Figure 2

A   C

B

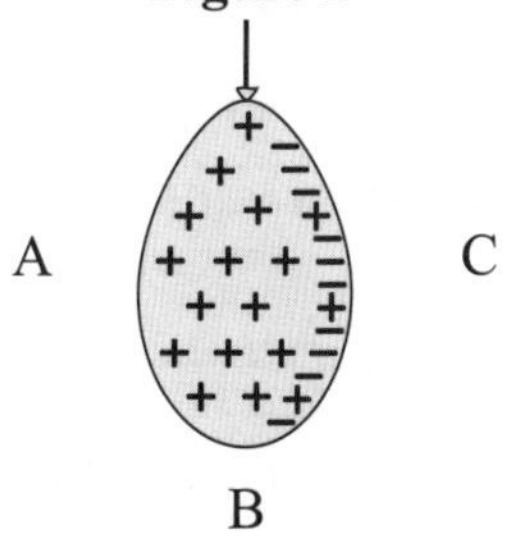

In which of the positions labelled A, B and C on the diagram did Jane hold the polythene rod?
Explain your answer.

[3 marks]

Revision Summary for Topic 2

You've made it through Topic 2 — time to put yourself through your paces and check it's all sunk in.
- Try these questions and tick off each one when you get it right.
- When you've done all the questions under a heading and are completely happy with it, tick it off.

Circuit Basics (p.40-43) ☑

1) Define current and state an equation that links current, charge and time, with units for each. ☑
2) What is meant by potential difference and resistance in a circuit? ☑
3) Draw the circuit symbols for: a cell, a lamp, a diode, a fuse and an LDR. ☑
4) What is the equation that links potential difference, current and resistance? ☑
5) What is an ohmic conductor? ☑
6) Explain how you would investigate how the length of a wire affects its resistance. ☑
7) Draw a circuit that could be used to investigate how the resistance of a filament bulb changes with the current through it. ☑
8) Name one linear component and one non-linear component. ☑

Circuit Devices and Types of Circuit (p.45-50) ☑

9) Explain how the resistance of an LDR varies with light intensity. ☑
10) What happens to the resistance of a thermistor as it gets colder? ☑
11) True or false? Potential difference is shared between components in a series circuit. ☑
12) True or false? The current is constant in a series circuit. ☑
13) True or false? The potential difference across each component connected in parallel is different. ☐
14) Explain why adding resistors in parallel decreases the total resistance of a circuit, but adding them in series increases the total resistance. ☑
15) Describe an experiment that could be carried out to investigate how adding resistors in series and parallel affects the total resistance of a circuit. ☑

Electricity in the Home (p.52-56) ☑

16) True or false? Mains supply electricity is an alternating current. ☐
17) What is the potential difference and the frequency of the UK mains supply? ☐
18) Name and give the colours of the three wires in a three-core cable. ☐
19) Give the potential differences for the three wires in a three-core mains cable. ☐
20) Explain why touching a live wire is dangerous. ☐
21) State three equations that can be used to calculate electrical power. ☑
22) What is the power rating of an appliance? ☑
23) Explain why electricity is transferred by the national grid at a high pd but low current. ☐
24) What are the functions of step-up and step-down transformers? ☐

Static Electricity and Electric Fields (p.58-60) ☑

25) How does rubbing of materials cause static electricity to build up? ☑
26) True or false? Two positive charges attract each other. ☑
27) In which direction do the arrows on electric field lines point? ☑
28) Using the concept of electric fields, explain how a build up of static electricity can cause a spark. ☐

Particle Model

The <u>particle model</u> is simpler than it sounds. It says that everything is made up of <u>lots of tiny particles</u> and describes how those particles behave in the three states of matter — <u>solids</u>, <u>liquids</u> and <u>gases</u>.

The **Particle Model** can Explain the **Three States of Matter**

1) In the <u>particle model</u>, you can think of the particles that make up matter as <u>tiny balls</u>. You can explain the ways that matter behaves in terms of how these tiny balls move, and the forces between them.

2) The <u>three states of matter</u> are <u>solid</u> (e.g. ice), <u>liquid</u> (e.g. water) and <u>gas</u> (e.g. water vapour). The <u>particles</u> of a substance in each state are the same — only the <u>arrangement</u> and <u>energy</u> of the particles are different.

Solids

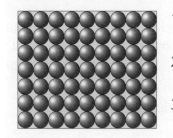

1) <u>Strong forces</u> of attraction hold the particles <u>close together</u> in a <u>fixed</u>, <u>regular</u> arrangement.

2) The particles don't have much <u>energy</u> so they <u>can</u> only <u>vibrate</u> about their <u>fixed</u> positions.

3) The <u>density</u> is generally <u>highest</u> in this state as the particles are <u>closest together</u>.

Liquids

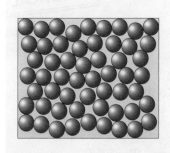

1) There are <u>weaker forces</u> of attraction between the particles.

2) The particles are <u>close together</u>, but can <u>move past each other</u>, and form <u>irregular</u> arrangements.

3) For any given substance, in the liquid state its particles will have <u>more energy</u> than in the solid state (but less energy than in the gas state).

4) They move in <u>random directions</u> at <u>low speeds</u>.

5) Liquids are generally <u>less dense</u> than solids.

Gases

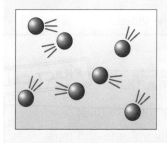

1) There are <u>almost no</u> forces of attraction between the particles.

2) For any given substance, in the gas state its particles will have <u>more energy</u> than in the solid state or the liquid state.

3) They are <u>free to move</u>, and travel in <u>random directions</u> and at <u>high speeds</u>.

4) Gases have <u>low</u> densities.

REVISION TIP

The higher their kinetic energy, the faster particles move...

Learn those diagrams above and make sure that you can describe the <u>arrangement</u> and <u>movement</u> of particles in solids, liquids and gases — it could earn you a few easy marks in the exam.

Density

Density tells you how much <u>mass</u> is packed into a given <u>volume</u> of space. You need to be able to work it out, as well as carry out <u>practicals</u> to work out the densities of different solids and liquids. Lucky you.

Density is a Measure of Compactness

Density is a measure of the '<u>compactness</u>' of a substance. It relates the <u>mass</u> of a substance to how much <u>space</u> it takes up (i.e. it's a substance's <u>mass</u> per <u>unit volume</u>). The <u>units</u> of <u>density</u> are <u>kg/m³</u> (the <u>mass</u> is in <u>kg</u> and the <u>volume</u> is in <u>m³</u>).

$$\text{Density (kg/m}^3) = \frac{\text{Mass (kg)}}{\text{Volume (m}^3)}$$

You might also see density given in g/cm^3. ($1 \, g/cm^3 = 1000 \, kg/m^3$)

1) The <u>density</u> of an <u>object</u> depends on what it's <u>made of</u>.

2) A <u>dense</u> material has its particles <u>packed tightly</u> together. The particles in a <u>less dense</u> material are more <u>spread out</u> — if you <u>compressed</u> the material, its particles would move <u>closer together</u>, and it would become <u>more dense</u>. (You <u>wouldn't</u> be changing its <u>mass</u>, but you <u>would</u> be <u>decreasing</u> its <u>volume</u>.)

3) This means that density varies between different <u>states of matter</u> (see previous page). <u>Solids</u> are generally <u>denser</u> than <u>liquids</u>, and <u>gases</u> are usually <u>less dense</u> than <u>liquids</u>.

You Need to be Able to Measure Density in Different Ways

To Find the Density of a Solid Object

PRACTICAL

1) Use a <u>balance</u> to measure its <u>mass</u> (see p.182).

2) For some solid shapes, you can find the <u>volume</u> using a <u>formula</u>. E.g. the volume of a cube is just width × height × length. *Make sure you know the formulas for the volumes of basic shapes.*

3) For a trickier shaped-solid, you can find its volume by <u>submerging</u> it in a <u>eureka can</u> filled with water. The water <u>displaced</u> by the object will be <u>transferred</u> to the <u>measuring cylinder</u>:

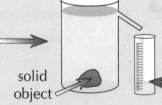

full eureka can

solid object

measuring cylinder

4) Record the <u>volume</u> of water in the measuring cylinder. This is the <u>volume</u> of the <u>object</u>.

5) Plug the object's <u>mass</u> and <u>volume</u> into the <u>formula</u> above to find its <u>density</u>.

To Find the Density of a Liquid

PRACTICAL

1) Place a <u>measuring cylinder</u> on a balance and <u>zero</u> the balance (see p.182).

2) Pour <u>10 ml</u> of the liquid into the measuring cylinder and record the liquid's <u>mass</u>.

3) Pour <u>another 10 ml</u> into the measuring cylinder and record the <u>total volume</u> and <u>mass</u>. Repeat this process until the measuring cylinder is <u>full</u>.

4) For each measurement, use the <u>formula</u> to find the <u>density</u>. (Remember that 1 ml = 1 cm³.)

5) Finally, take an <u>average</u> of your calculated densities to get an accurate value for the <u>density</u> of the <u>liquid</u>.

Internal Energy and Changes of State

This page is all about heating things. Take a look at your specific heat capacity notes (p.20) before you start — you need to understand it and be able to be able to use $\Delta E = mc\Delta\theta$ for this topic too I'm afraid.

Internal Energy is Stored by the Particles That Make Up a System

1) The particles in a system vibrate or move around — they have energy in their kinetic energy stores.

2) They also have energy in their potential energy stores due to their positions — don't worry about this.

3) The energy stored in a system is stored by its particles (atoms and molecules). The internal energy of a system is the total energy that its particles have in their kinetic and potential energy stores.

4) Heating the system transfers energy to its particles (they gain energy in their kinetic stores and move faster), increasing the internal energy.

5) This leads to a change in temperature or a change in state. If the temperature changes, the size of the change depends on the mass of the substance, what it's made of (its specific heat capacity) and the energy input. Make sure you remember all of the stuff on specific heat capacity from p.20, particularly how to use the formula.

6) A change in state occurs if the substance is heated enough — the particles will have enough energy in their kinetic energy stores to break the bonds holding them together.

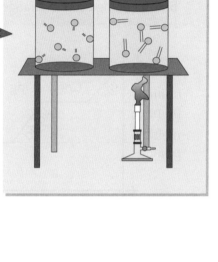

A Change of State Conserves Mass

1) When you heat a liquid, it boils (or evaporates) and becomes a gas. When you heat a solid, it melts and becomes a liquid. These are both changes of state.

2) The state can also change due to cooling. The particles lose energy and form bonds.

3) The changes of state are:

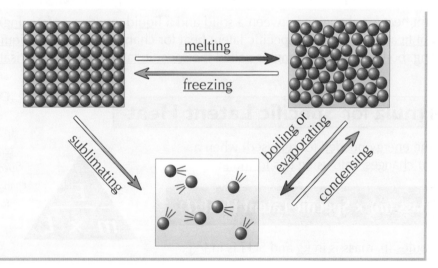

4) A change of state is a physical change (rather than a chemical change). This means you don't end up with a new substance — it's the same substance as you started with, just in a different form.

5) If you reverse a change of state (e.g. freeze a substance that has been melted), the substance will return to its original form and get back its original properties.

6) The number of particles doesn't change — they're just arranged differently. This means mass is conserved — none of it is lost when the substance changes state.

Specific Latent Heat

The energy needed to change the state of a substance is called latent heat. This is exciting stuff I tell you...

A **Change of State Requires Energy**

When a substance is melting or boiling, you're still putting in energy and so increasing the internal energy, but the energy's used for breaking bonds between particles rather than raising the temperature. There are flat spots on the heating graph where energy is being transferred by heating but not being used to change the temperature.

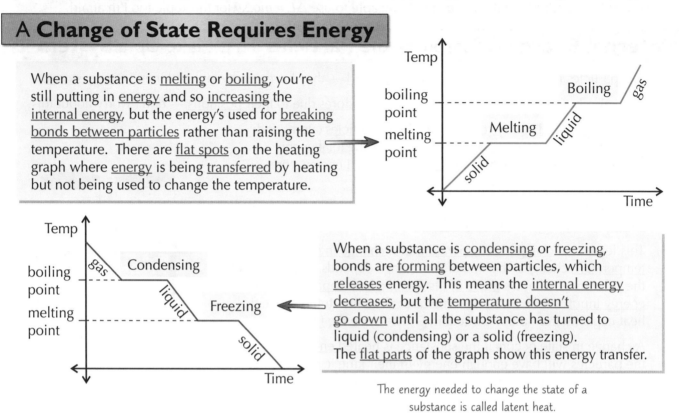

When a substance is condensing or freezing, bonds are forming between particles, which releases energy. This means the internal energy decreases, but the temperature doesn't go down until all the substance has turned to liquid (condensing) or a solid (freezing). The flat parts of the graph show this energy transfer.

The energy needed to change the state of a substance is called latent heat.

Specific Latent Heat is the **Energy Needed** to **Change State**

1) The specific latent heat (SLH) of a substance is the amount of energy needed to change 1 kg of it from one state to another without changing its temperature.

2) For cooling, specific latent heat is the energy released by a change in state.

3) Specific latent heat is different for different materials, and for changing between different states.

4) The specific latent heat for changing between a solid and a liquid (melting or freezing) is called the specific latent heat of fusion. The specific latent heat for changing between a liquid and a gas (evaporating, boiling or condensing) is called the specific latent heat of vaporisation.

There's a **Formula** for **Specific Latent Heat**

You can work out the energy needed (or released) when a substance of mass m changes state using this formula:

Energy (E) = Mass (m) × Specific Latent Heat (L)

Don't get confused with specific heat capacity (p.20), which relates to a temperature rise of 1 °C. Specific latent heat is about changes of state where there's no temperature change.

Energy is given in joules (J), mass is in kg and SLH is in J/kg.

EXAMPLE: **The specific latent heat of vaporisation for water (boiling) is 2 260 000 J/kg. How much energy is needed to completely boil 1.50 kg of water at 100 °C?**

1) Just plug the numbers into the formula.

$E = mL$
$= 1.50 × 2 260 000$

2) The units are joules because it's energy.

$= 3 390 000$ J

Particle Motion in Gases

The particle model helps explain how temperature, pressure, volume and energy in kinetic stores are all related. And this page is here to explain it all to you. I bet you're just itching to find out more...

Average Energy in Kinetic Stores is Related to Temperature

1) The particles in a gas are constantly moving with random directions and speeds. If you increase the temperature of a gas, you transfer energy into the kinetic energy stores of its particles (see page 17 for more on energy stores).

2) The temperature of a gas is related to the average energy in the kinetic energy stores of the particles in the gas. The higher the temperature, the higher the average energy.

3) So as you increase the temperature of a gas, the average speed of its particles increases. This is because the energy in the particles' kinetic energy stores is $\frac{1}{2}mv^2$ — p.19.

Colliding Gas Particles Create Pressure

1) As gas particles move about at high speeds, they bang into each other and whatever else happens to get in the way. When they collide with something, they exert a force (and so a pressure — p.100) on it. In a sealed container, the outward gas pressure is the total force exerted by all of the particles in the gas on a unit area of the container walls.

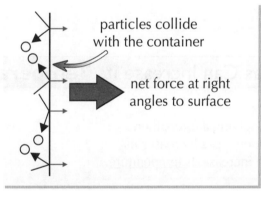

particles collide with the container

net force at right angles to surface

2) Faster particles and more frequent collisions both lead to an increase in net force, and so gas pressure.

3) One way of increasing the speed of the particles is to heat them up. If volume is kept constant, then increasing temperature will increase the speed of the particles, and so the pressure.

4) Alternatively, if temperature is constant, increasing the volume of a gas means the particles get more spread out and hit the walls of the container less often. The gas pressure decreases.

5) Pressure and volume are inversely proportional — at a constant temperature, when volume goes up, pressure goes down (and when volume decreases, pressure increases).

6) For a gas of fixed mass at a constant temperature, the relationship is:

$$pV = \text{constant}$$

p = pressure, in pascals (Pa)
V = volume (m³)

Higher temperatures mean higher average energies in kinetic stores...

The particle model can be used to explain what happens when you change the temperature of a gas which is kept at a constant volume. Have a look back on page 63 for more about the particle model.

Pressure of Gases

So you've had a whole page on <u>particle motion in gases</u> and even an equation to boot. You're not done yet though. On the bright side, there is mention of <u>balloons</u> on this page, so it's sort of like a party.

A Change in **Pressure** can Cause a Change in **Volume**

1) The <u>pressure</u> of a gas causes a <u>net outwards force</u> at right angles to the surface of its container.

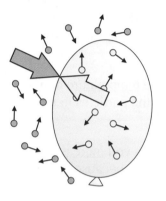

2) There is also a force on the <u>outside</u> of the container due to the pressure of the gas <u>around it</u>.

3) If a container can easily <u>change its size</u> (e.g. a balloon), then any change in these pressures will cause the container to <u>compress</u> or <u>expand</u>, due to the overall force.

> E.g. if a <u>helium balloon</u> is released, it rises. Atmospheric pressure <u>decreases</u> with height (p.102), so the pressure <u>outside</u> the balloon <u>decreases</u>. This causes the balloon to <u>expand</u> until the pressure inside <u>drops</u> to the same as the atmospheric pressure.

Doing **Work** on a Gas Can Increase its **Temperature**

1) If you <u>transfer energy</u> by applying a <u>force</u>, then you do <u>work</u>. Doing work on a gas increases its <u>internal energy</u>, which can increase its <u>temperature</u>.

There's more about doing work on p.90.

2) You can do work on a gas <u>mechanically</u>, e.g. with a <u>bike pump</u>. The gas <u>applies pressure</u> to the <u>plunger</u> of the pump and so exerts a <u>force</u> on it. Work has to be done <u>against this force</u> to push down the plunger.

3) This transfers energy to the <u>kinetic energy stores</u> of the gas particles, increasing the <u>temperature</u>. If the pump is connected to a <u>tyre</u>, you should feel the tyre <u>getting warmer</u>.

Gases can be compressed or expanded by pressure changes...

You should be able to explain how, in any given situation (including that example of the bicycle pump above), doing <u>work</u> on a <u>gas</u> that's <u>enclosed</u> will lead to an <u>increase</u> in <u>temperature</u> of the gas. It will help if you really understand work too. Work's just the transfer of energy by a force.

Topic 3 — Particle Model of Matter

Warm-Up & Exam Questions

Once you think you've got to grips with everything in this topic, all the way through from the particle model to that stuff about gases, it's time to test yourself with these questions. Let's see how you get on.

Warm-Up Questions

1) Describe the particles in a liquid in terms of their arrangement, energy and movement.
2) What is density a measure of?
3) How does cooling a system affect its internal energy?
4) What is the specific latent heat of vaporisation?
5) What are the units of specific latent heat?
6) Describe and explain how the pressure inside and outside a helium balloon would change if it was allowed to float up in the atmosphere.

Exam Questions

1   Substances can exist in different states of matter. **Grade 4-6**

1.1   Describe the arrangement and movement of the particles in a solid.

[2 marks]

If a substance is heated to a certain temperature it can change from a solid to a liquid.

1.2   Give the name of this process.

[1 mark]

1.3   If a liquid is heated to a certain temperature it starts to boil and become a gas.
Name the other process that causes a liquid to start to become a gas.

[1 mark]

PRACTICAL

2   A student has a collection of metal toy soldiers of different sizes made from the same metal. **Grade 6-7**

2.1   Which of the following statements about the toy soldiers is true? Tick **one** box.

☐ The masses and densities of each of the toy soldiers are the same.

☐ The masses of each of the toy soldiers are the same, but their densities may vary.

☐ The densities of each of the toy soldiers are the same, but their masses may vary.

☐ The densities and masses of each toy soldier may vary.

[1 mark]

The student wants to measure the density of one of the toy soldiers.
He has a eureka can, a measuring cylinder, a mass balance and some water.

2.2   State the **two** quantities the student must measure in order to calculate the density of the toy soldier.

[2 marks]

2.3*   Describe the steps the student could take to find the density
of the toy soldier using the equipment he has.

[6 marks]

Exam Questions

3   The following question is about specific latent heat. (Grade 6-7)

3.1   What is the name given to the specific latent heat of a
substance when it's changing between a solid and a liquid?

[1 mark]

3.2   The energy required to convert 40.8 g of liquid methanol to gaseous methanol is 47 100 J.
Calculate the specific latent heat of vaporisation of methanol.
Use the correct equation from the Physics Equation Sheet on the inside back cover.
Give your answer in J/kg and to 3 significant figures.

[3 marks]

3.3   The energy used to change liquid methanol to gaseous methanol
was supplied by heating the system.
How does the internal energy of the system change as the system is heated? Explain your answer.

[2 marks]

3.4   Explain, using the particle model, what happens when methanol
is heated so that it changes from a liquid to a gas.

[2 marks]

4   The movement of particles in a gas can be described using the particle model. (Grade 6-7)

4.1   Describe the particles in a gas using the particle model.
In your answer, you should refer to the arrangement, energy and movement of the particles.

[3 marks]

A scientist is calculating the density of a gas in a sealed, rigid container.

4.2   Use the particle model to explain how gas particles create pressure in a sealed container.

[2 marks]

4.3   State the equation that links density with mass and volume.

[1 mark]

4.4   A certain gas had a mass of 8.2 g and a volume of 6.69 cm^3.
Calculate the density of the gas.
Give your answer to an appropriate number of significant figures.

[4 marks]

4.5   The scientist heats the container.
What happens to the pressure of the gas within the container?
Explain your answer using the particle model.

[3 marks]

5   A student is doing an investigation into the masses of different materials of different densities.
A cube has edges of length 1.5 cm and a density of 3500 kg/m^3.
Calculate the mass of the cube. (Grade 6-7)

[5 marks]

6   A sealed balloon contains 0.034 m^3 of gas at a pressure of 98 kPa.
The balloon is compressed to 0.031 m^3. The temperature of the air inside it remains constant. (Grade 7-9)
Calculate the air pressure inside the balloon after the compression.

[3 marks]

Revision Summary for Topic 3

Don't let all that stuff about gas pressures in <u>Topic 3</u> put too much pressure on you. Try these questions to see whether you've really got to grips with all of that stuff about <u>states of matter</u> and the <u>particle model</u>.
*   Try these questions and <u>tick off each one</u> when you <u>get it right</u>.
*   When you've done <u>all the questions</u> under a heading, and are <u>completely happy</u> with it, tick it off.

Particle Model (p.63) ☑
1) What are the three states of matter?
2) For each state of matter, describe the arrangement and movement of the particles.

Density (p.64) ☑
3) What is the formula for density?
4) Give an example of the possible units of density.
5) Describe how you could find the volume of an irregular solid object.
6) Briefly describe an experiment to find the density of a liquid.

Internal Energy and Changes of State (p.65) ☑
7) What is internal energy?
8) What happens to the particles in a substance when that substance is heated?
9) Name the six changes of state.
10) Is a change of state a physical change or a chemical change?
11) True or false? Mass stays the same when a substance changes state.

Specific Latent Heat (p.66) ☑
12) Explain the cause of the flat sections on a graph of temperature against time for a substance being heated.
13) Sketch a graph of temperature against time for a gas being cooled. Your graph should show the points that the gas turns into a liquid and that the liquid turns into a solid.
14) Define specific latent heat.
15) What is meant by the term 'specific heat of fusion'?

Particle Motion in Gases (p.67) ☑
16) Explain how a gas in a sealed container exerts a pressure on the walls of the container.
17) A sealed container of gas is kept at a constant temperature. The volume of the container is increased. What happens to the pressure of the gas? Explain why.
18) For a fixed mass of gas at a constant temperature, what is the relationship between pressure and volume?
19) True or false? For a gas at constant temperature, increasing the volume of the gas will also increase its pressure.

Pressure in Gases (p.68) ☑
20) A balloon containing a fixed mass of helium gas is moved from an area of high atmospheric pressure to one of low atmospheric pressure. What will happen to the volume of helium in the balloon?
21) Explain why blowing up a football with a pump causes the ball to warm up.

Developing the Model of the Atom

All this started with a Greek chap called Democritus in the 5th Century BC. He thought that <u>all matter</u>, whatever it was, was made up of <u>identical</u> lumps called "atomos". And that's as far as it got until the 1800s...

The **Plum Pudding** Model was **Replaced** with the **Nuclear** Model

1) In 1804 <u>John Dalton</u> agreed with Democritus that matter was made up of <u>tiny spheres</u> ("atoms") that couldn't be broken up, but he reckoned that <u>each element</u> was made up of a <u>different type</u> of "atom".

2) Nearly 100 years later, J. J. Thomson discovered particles called <u>electrons</u> that could be <u>removed</u> from atoms. So Dalton's theory wasn't quite right (atoms could be broken up). Thomson suggested that atoms were <u>spheres of positive charge</u> with tiny negative electrons <u>stuck in them</u> like the fruit in a <u>plum pudding</u> — the <u>plum pudding model</u>.

3) That "plum pudding" theory didn't last though... In 1909, scientists in <u>Rutherford's</u> lab tried firing a beam of <u>alpha particles</u> (see p.75) at <u>thin gold foil</u> — this was the <u>alpha scattering experiment</u>. From the plum pudding model, they expected the particles to <u>pass straight through</u> the gold sheet, or only be <u>slightly deflected</u>.

4) But although most of the particles did go <u>straight through</u> the sheet, some were deflected more than they had expected, and a few were <u>deflected back</u> the way they had come — something the plum pudding model <u>couldn't explain</u>.

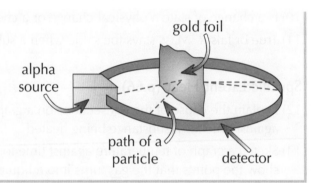

5) Because a few alpha particles were deflected back, the scientists realised that <u>most of the mass</u> of the atom was concentrated at the <u>centre</u> in a <u>tiny nucleus</u>. This nucleus must also have a <u>positive charge</u>, since it repelled the positive alpha particles.

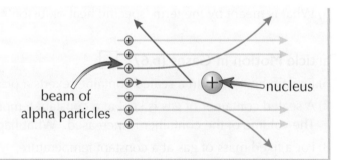

6) They also realised that because <u>nearly all</u> the alpha particles passed <u>straight through</u>, most of an atom is just <u>empty space</u>. This was the <u>first nuclear model</u> of the atom.

The gold foil experiment helped adapt the model of the atom...

Rutherford and his lab of scientists made a <u>hypothesis</u>, did an <u>investigation</u> and then <u>analysed</u> the data they got from it. By doing this, they showed that the plum pudding model of the atom must be wrong, so it was <u>changed</u>. This is a great example of the <u>scientific method</u> (see page 1) in action.

Developing the Model of the Atom

Rutherford and Marsden's model of the atom was a big leap forwards, but that's not the end of the story...

Bohr **Refined** Rutherford's **Nuclear Model** of the Atom

1) The nuclear model that resulted from the alpha particle scattering experiment was a positively charged nucleus surrounded by a cloud of negative electrons.

2) Niels Bohr said that electrons orbiting the nucleus do so at certain distances called energy levels. His theoretical calculations agreed with experimental data.

3) Evidence from further experiments changed the model to have a nucleus made up of a group of particles (protons) which all had the same positive charge that added up to the overall charge of the nucleus.

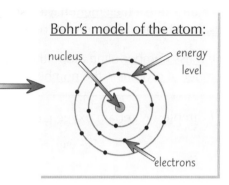

Bohr's model of the atom:

nucleus · energy level · electrons

4) About 20 years after the idea of a nucleus was accepted, in 1932, James Chadwick proved the existence of the neutron, which explained the imbalance between the atomic and mass numbers (see next page).

Our **Current Model** of the **Atom**

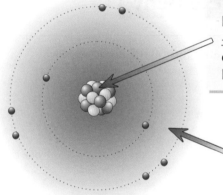

The nucleus is tiny but it makes up most of the mass of the atom. It contains protons (which are positively charged — they have a +1 relative charge) and neutrons (which are neutral, with a relative charge of 0) — which gives it an overall positive charge. Its radius is about 10 000 times smaller than the radius of the atom.

The rest of the atom is mostly empty space. Negative electrons (which have a relative charge of –1) whizz round the outside of the nucleus really fast. They give the atom its overall size — the radius of an atom is about 1×10^{-10} m.

We're currently pretty happy with this model, but there's no saying it won't change. Just like for the plum pudding model, new experiments sometimes mean we have to change or completely get rid of current models.

Number of Protons **Equals** Number of Electrons

1) In atoms, the number of protons = the number of electrons, as protons and electrons have an equal but opposite charge and atoms have no overall charge.

2) Electrons in energy levels can move within (or sometimes leave) the atom. If they gain energy by absorbing EM radiation (p.136) they move to a higher energy level, further from the nucleus. If they release EM radiation, they move to a lower energy level that is closer to the nucleus. If one or more outer electrons leaves the atom, the atom becomes a positively charged ion.

WORKING SCIENTIFICALLY

The model of the atom has developed over time...

Due to lots of scientists doing lots of experiments, we now have a better idea of what the atom's really like. We now know about the particles in atoms — protons, neutrons and electrons.

Isotopes

Isotopes of an element look pretty similar, but watch out — they have different numbers of neutrons.

Atoms of the Same Element have the Same Number of Protons

1) All atoms of each element have a set number of protons (so each nucleus has a given positive charge). The number of protons in an atom is its atomic number.

2) The mass number of an atom (the mass of the nucleus) is the number of protons + the number of neutrons in its nucleus.

> Example: A certain oxygen atom has the chemical symbol — $^{16}_{8}O$.
>
> Mass number ⟶ 16 O ⟵ Element symbol (oxygen)
> Atomic number ⟶ 8
>
> • Oxygen has an atomic number of 8, this means all oxygen atoms have 8 protons.
> • This atom of oxygen has a mass number of 16.
>   Since it has 8 protons, it must have $16 - 8 = 8$ neutrons.

Isotopes are Different Forms of the Same Element

1) Isotopes of an element are atoms with the same number of protons (the same atomic number, and so the same charge on the nucleus) but a different number of neutrons (a different mass number).

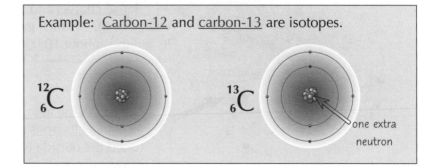

Example: Carbon-12 and carbon-13 are isotopes.

$^{12}_{6}C$ $^{13}_{6}C$ one extra neutron

2) All elements have different isotopes, but there are usually only one or two stable ones.

3) The other unstable isotopes tend to decay into other elements and give out radiation as they try to become more stable. This process is called radioactive decay.

4) Radioactive substances spit out one or more types of ionising radiation from their nucleus — the ones you need to know are alpha, beta and gamma radiation (see next page).

5) They can also release neutrons (n) when they decay.

6) Ionising radiation is radiation that knocks electrons off atoms, creating positive ions. The ionising power of a radiation source is how easily it can do this.

Ionising Radiation

There are three types of ionising radiation you need to know about — these are <u>alpha</u>, <u>beta</u> and <u>gamma</u>.

Alpha Particles are Helium Nuclei

1) Alpha radiation is when an <u>alpha particle</u> (α) is emitted from the nucleus. An α-particle is <u>two neutrons</u> and <u>two protons</u> (like a <u>helium nucleus</u>).

2) They <u>don't</u> penetrate very far into materials and are <u>stopped quickly</u> — they can only travel a <u>few cm in air</u> and are <u>absorbed</u> by a sheet of <u>paper</u>.

3) Because of their size they are <u>strongly ionising</u>.

4) Alpha radiation has <u>applications</u> in the home:

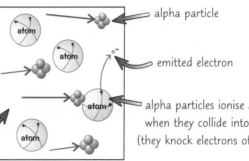

alpha particle

emitted electron

alpha particles ionise atoms when they collide into them (they knock electrons off them)

> Alpha radiation is used in smoke detectors — it <u>ionises</u> air particles, causing a <u>current</u> to flow. If there is smoke in the air, it <u>binds</u> to the ions — meaning the current stops and the alarm sounds.

Beta Particles are High-Speed Electrons

1) A <u>beta particle</u> (β) is simply a fast-moving <u>electron</u> released by the nucleus. Beta particles have virtually <u>no mass</u> and a charge of –1.

2) They are <u>moderately ionising</u> (see right).

3) They also <u>penetrate moderately</u> far into materials before colliding and have a <u>range in air</u> of a <u>few metres</u>. They are <u>absorbed</u> by a sheet of <u>aluminium</u> (around <u>5 mm thick</u>).

4) For every <u>beta particle</u> emitted, a neutron in the nucleus has turned into a <u>proton</u>.

5) Beta radiation can be useful due to the fact that it's <u>moderately penetrating</u>:

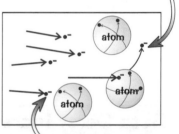

emitted electron

beta particle

> <u>Beta emitters</u> are used to test the <u>thickness</u> of <u>sheets of metal</u>, as the particles are not immediately absorbed by the material like alpha radiation would be and do not penetrate as far as gamma rays. Therefore, slight variations in thickness affect the <u>amount of radiation</u> passing through the sheet.

Gamma Rays are EM Waves with a Short Wavelength

1) <u>Gamma rays</u> (γ) are waves of <u>electromagnetic radiation</u> (p.136) released by the nucleus.

2) They <u>penetrate far into materials</u> without being stopped and will travel a <u>long distance</u> through <u>air</u>.

3) This means they are <u>weakly</u> ionising because they tend to <u>pass through</u> rather than collide with atoms. Eventually they <u>hit something</u> and do <u>damage</u>.

4) They can be <u>absorbed</u> by thick sheets of <u>lead</u> or metres of <u>concrete</u>.

Uses of gamma rays are on p.83 and p.139.

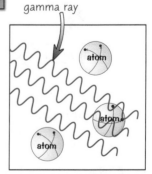

gamma ray

Alpha particles are more ionising than beta particles...

...and <u>beta particles</u> are <u>more ionising</u> than <u>gamma rays</u>. Make sure you've got that clearly memorised, as well as what makes up each type of <u>radiation</u>, as this isn't the last you'll see of this stuff. No siree.

Nuclear Equations

Nuclear equations show radioactive decay and once you get the hang of them they're dead easy. Get going.

Mass and Atomic Numbers Have to Balance

1) Nuclear equations are a way of showing radioactive decay by using element symbols (p.74). They're written in the form: atom before decay → atom(s) after decay + radiation emitted.

2) There is one golden rule to remember: the total mass and atomic numbers must be equal on both sides.

Alpha Decay Decreases the Charge and Mass of the Nucleus

1) Remember, alpha particles are made up of two protons and two neutrons. So when an atom emits an alpha particle, its atomic number reduces by 2 and its mass number reduces by 4.

2) A proton is positively charged and a neutron is neutral, so the charge of the nucleus decreases.

3) In nuclear equations, an alpha particle can be written as a helium nucleus: $^{4}_{2}\text{He}$.

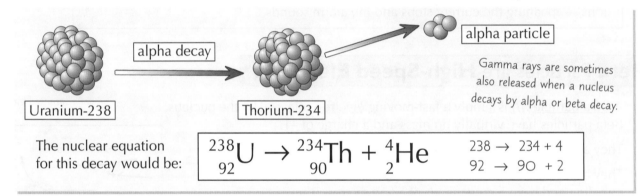

Uranium-238 → alpha decay → Thorium-234 + alpha particle

Gamma rays are sometimes also released when a nucleus decays by alpha or beta decay.

The nuclear equation for this decay would be:

$$^{238}_{92}\text{U} \rightarrow\ ^{234}_{90}\text{Th} +\ ^{4}_{2}\text{He}$$

$238 \rightarrow 234 + 4$
$92 \rightarrow 90 + 2$

Beta Decay Increases the Charge of the Nucleus

1) When beta decay occurs, a neutron in the nucleus turns into a proton and releases a fast-moving electron (the beta particle).

2) The number of protons in the nucleus has increased by 1. This increases the positive charge of the nucleus (the atomic number).

3) Because the nucleus has lost a neutron and gained a proton during beta decay, the mass of the nucleus doesn't change (protons and neutrons have the same mass).

4) A beta particle is written as $^{0}_{-1}\text{e}$ in nuclear equations.

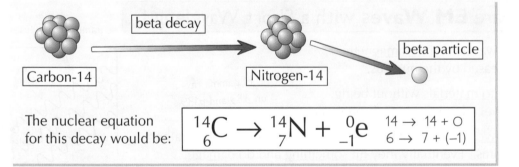

Carbon-14 → beta decay → Nitrogen-14 + beta particle

In both alpha and beta emissions, a new element will be formed, as the number of protons (atomic number) changes.

The nuclear equation for this decay would be:

$$^{14}_{6}\text{C} \rightarrow\ ^{14}_{7}\text{N} +\ ^{0}_{-1}\text{e}$$

$14 \rightarrow 14 + 0$
$6 \rightarrow 7 + (-1)$

Gamma Rays Don't Change the Charge or Mass of the Nucleus

1) Gamma rays are a way of getting rid of excess energy from a nucleus.

2) This means that there is no change to the atomic mass or atomic number of the atom.

3) In nuclear equations, gamma radiation is written as $^{0}_{0}\gamma$.

Topic 4 — Atomic Structure

Half-Life

How quickly <u>unstable nuclei</u> decay is measured using <u>activity</u> and <u>half-life</u> — two very important terms.

Radioactivity is a Totally Random Process

1) Radioactive substances give out <u>radiation</u> from the nuclei of their atoms — <u>no matter what</u>.

2) This radiation can be measured with a <u>Geiger-Muller tube and counter</u>, which records the <u>count-rate</u> — the number of radiation counts reaching it per second.

3) Radioactive decay is entirely <u>random</u>. So you <u>can't predict</u> exactly <u>which</u> nucleus in a sample will decay next, or <u>when</u> any one of them will decay.

4) But you <u>can</u> find out the <u>time</u> it takes for the <u>amount of radiation</u> emitted by a source to <u>halve</u>, this is known as the <u>half-life</u>. It can be used to make <u>predictions</u> about radioactive sources, even though their decays are <u>random</u>.

5) Half-life can be used to find the <u>rate</u> at which a source decays — its <u>ACTIVITY</u>. Activity is measured in <u>becquerels, Bq</u> (where 1 Bq is <u>1 decay per second</u>).

The Radioactivity of a Source Decreases Over Time

1) Each time a radioactive nucleus <u>decays</u> to become a <u>stable nucleus</u>, the activity <u>as a whole</u> will <u>decrease</u>. (<u>Older</u> sources emit <u>less</u> radiation.)

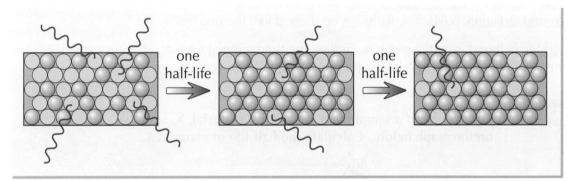

2) For <u>some</u> isotopes it takes <u>just a few hours</u> before nearly all the unstable nuclei have <u>decayed</u>, whilst others last for <u>millions of years</u>.

3) The problem with trying to <u>measure</u> this is that <u>the activity never reaches zero</u>, which is why we have to use the idea of <u>half-life</u> to measure how quickly the activity <u>drops off</u>.

The <u>half-life</u> is the time taken for the <u>number of radioactive nuclei</u> in an isotope to <u>halve</u>.

4) Half-life can also be described as the time taken for the <u>activity</u>, and so count-rate, to <u>halve</u>.

5) A <u>short half-life</u> means the <u>activity falls quickly</u>, because the nuclei are very <u>unstable</u> and <u>rapidly decay</u>. Sources with a short half-life can be <u>dangerous</u> because of the <u>high</u> amount of radiation they emit at the start, but they <u>quickly</u> become <u>safe</u>.

6) A <u>long half-life</u> means the activity <u>falls more slowly</u> because <u>most</u> of the nuclei don't decay <u>for a long time</u> — the source just sits there, releasing <u>small</u> amounts of radiation for a <u>long time</u>. This can be dangerous because <u>nearby areas</u> are <u>exposed</u> to radiation for (<u>millions</u> of) <u>years</u>.

Different substances have different half-lives...

Some substances take a <u>long time</u> to <u>decay</u>, giving them a <u>long half-life</u>, while others decay in the blink of an eye. For example, neodymium-144 has a half-life of 2 million billion years, while helium-5 has a half-life of 7.6×10^{-22} seconds. That's 0.00000000000000000000076 seconds. Pretty speedy eh?

Half-Life

You learnt all about what half-life is on the last page, but now it's time to find out how to calculate it. Fortunately, there's a pretty simple method you can use that involves an activity-time graph.

The Radioactivity of a Source Decreases Over Time

You might be asked to give the decline of activity or count-rate after a certain number of half-lives as a percentage of the original activity, like this:

EXAMPLE: **The initial activity of a sample is 640 Bq. Calculate the final activity as a percentage of the initial activity after two half-lives.**

1) Find the activity after each half-life.

2) Now divide the final activity by the initial activity, then multiply by 100 to make it a percentage.

1 half-life: $640 ÷ 2 = 320$
2 half-lives: $320 ÷ 2 = 160$
$(160 ÷ 640) × 100$
$= 0.25 × 100$
$= 25\%$

Always double check what the question is asking for — it may want a fraction, ratio or a percentage.

Finding the Half-Life of a Sample using a Graph

1) If you plot a graph of activity against time (taking into account background radiation, p.81), it will always be shaped like the one below.

2) The half-life is found from the graph by finding the time interval on the bottom axis corresponding to a halving of the activity on the vertical axis. Easy.

EXAMPLE: **The activity of a sample of a radioactive material, X, is shown on the graph below. Calculate the half-life of material X.**

1) Read the initial activity off the graph. This is the activity when time = 0.

2) Divide the initial activity by 2 to find the value of half the initial activity.

$80 ÷ 2 = 40$

3) Find this value on the y-axis and read along horizontally to the curve.

4) Then read down from the curve at this point to find the half-life.

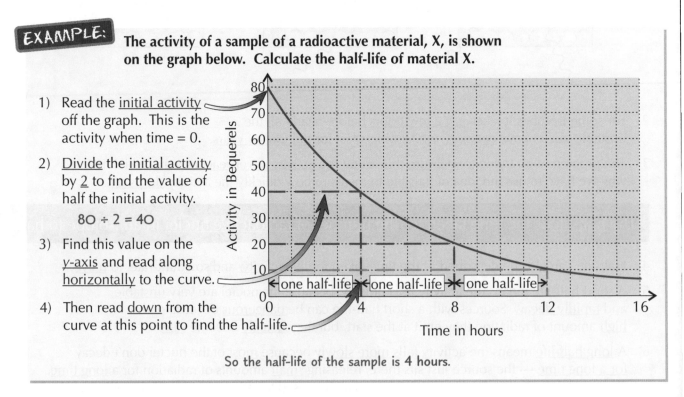

So the half-life of the sample is 4 hours.

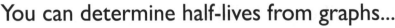

You can determine half-lives from graphs...

MATHS TIP Make sure you can use graphs like the one above to work out half-lives. All you've got to do is read off the initial activity from the y-axis, then work out what half this activity would be by dividing by two. Then, just read off the time from the x-axis for this value, which is one half-life.

Warm-Up & Exam Questions

Atoms may be tiny, but you could bag some big marks in your exams if you know them inside-out. Here are some questions to check just how great your understanding of atoms and radiation really is...

Warm-Up Questions

1) Describe our current, nuclear model of the atom.
2) Give the definition of the term 'isotope'.
3) Which are the most ionising — alpha particles or gamma rays?
4) Outline why beta emitters, rather than alpha or gamma emitters, are used to test the thickness of sheets of metal.
5) Name the type of nuclear radiation, the particles of which are electrons.
6) Name the type of nuclear radiation that is an electromagnetic wave.
7) Radioactive substances with short half-lives can be initially very dangerous. Explain why.

Exam Questions

1   Alpha, beta and gamma radiation sources were used to direct radiation at thin sheets of paper and aluminium. A detector was used to measure where radiation had passed through the sheets. The results are shown in **Figure 1**.

Grade 4-6

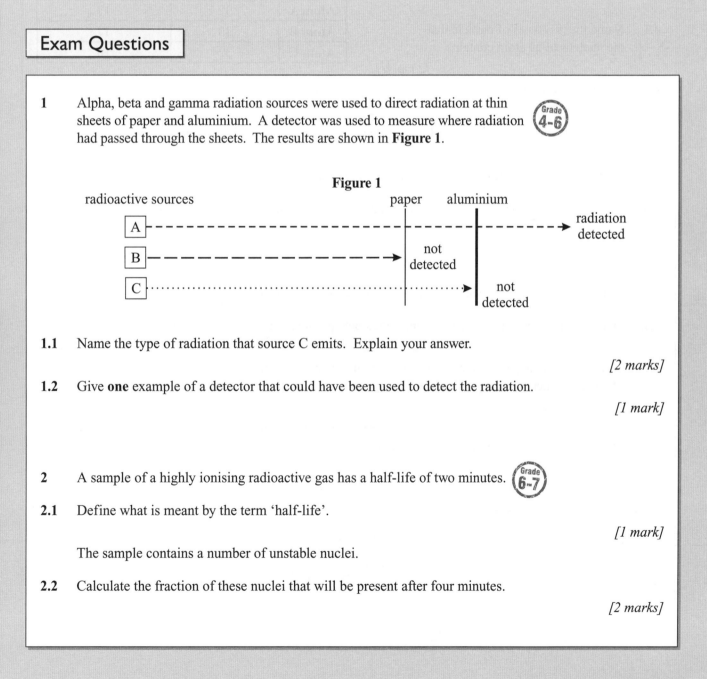

Figure 1

1.1   Name the type of radiation that source C emits. Explain your answer.

[2 marks]

1.2   Give **one** example of a detector that could have been used to detect the radiation.

[1 mark]

2   A sample of a highly ionising radioactive gas has a half-life of two minutes.

Grade 6-7

2.1   Define what is meant by the term 'half-life'.

[1 mark]

The sample contains a number of unstable nuclei.

2.2   Calculate the fraction of these nuclei that will be present after four minutes.

[2 marks]

Exam Questions

3   A radioactive isotope sample has a half-life of 40 seconds. *(Grade 6-7)*
The initial activity of the sample is 8000 Bq.

3.1   Calculate the activity after 2 minutes. Give your answer in becquerels.

[2 marks]

3.2   After how many half-lives will the activity have fallen to 250 Bq?

[2 marks]

3.3   The radioactive source is left until its activity falls to 100 Bq.
Calculate the final activity as a percentage of the initial activity.

[2 marks]

4   **Table 1** contains information *(Grade 6-7)*
about three atoms.

4.1   Name the two types of particle that
the nucleus of an atom contains.

[2 marks]

Table 1

	Mass number	Atomic number
Atom A	32	17
Atom B	33	17
Atom C	32	16

4.2   Define the term 'mass number' in the context of atoms.

[1 mark]

4.3   Which of the two atoms in **Table 1** are isotopes of the same element? Explain your answer.

[2 marks]

Alpha and beta particles are deflected in electric fields.

4.4   Suggest why alpha and beta particles are deflected in opposite directions.

[1 mark]

5   Nuclear equations show what is produced when unstable nuclei decay. *(Grade 7-9)*

5.1   Draw a symbol that can be used to represent a beta particle in a nuclear equation.

[1 mark]

5.2   Describe what happens to the atomic number and the mass number
of an atom when it undergoes beta decay.

[2 marks]

5.3   Describe what happens to the atomic number and the mass number
of an atom when it undergoes gamma decay.

[2 marks]

5.4   Complete the nuclear equation, shown in **Figure 2**, which shows
a polonium isotope decaying by alpha emission.

Figure 2

$$_{84}^{\ \ \ldots}\text{Po} \longrightarrow _{\ \ldots}^{205}\text{Pb} + _{\ \ldots}^{\ \ldots}\text{He}$$

[3 marks]

Background Radiation

Forget love — <u>radiation</u> is <u>all around</u>. Don't panic too much though, it's usually a pretty <u>small amount</u>.

Background Radiation Comes From Many Sources

<u>Background radiation</u> is the <u>low-level</u> radiation that's around us <u>all the time</u>. You should always <u>measure</u> and <u>subtract</u> the background radiation from your results (to avoid systematic errors, p.7). It comes from:

1) Radioactivity of naturally occurring <u>unstable isotopes</u> which are <u>all around us</u> — in the <u>air</u>, in <u>food</u>, in <u>building materials</u> and in the <u>rocks</u> under our feet.

2) Radiation from <u>space</u>, which is known as <u>cosmic rays</u>. These come mostly from the <u>Sun</u>. Luckily, the Earth's <u>atmosphere protects</u> us from much of this radiation.

3) Radiation due to <u>human activity</u>, e.g. <u>fallout</u> from <u>nuclear explosions</u> or <u>nuclear waste</u>. But this represents a <u>tiny</u> proportion of the total background radiation.

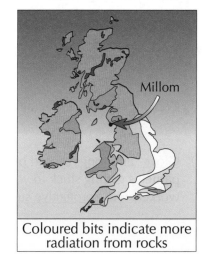

Coloured bits indicate more radiation from rocks

The <u>radiation dose</u> tells you the <u>risk of harm</u> to body tissues due to exposure to radiation. It's measured in <u>sieverts</u> (Sv) (p.140). The dose from background radiation is <u>small</u>, so <u>millisieverts</u> are often used to measure it (<u>1 Sv = 1000 mSv</u>). Your radiation dose is affected by <u>where you live</u> and whether you have a <u>job</u> that involves <u>radiation</u>. Other factors, such as having <u>x-rays</u> taken, also affect your radiation dose (see p.139).

Exposure to Radiation is called Irradiation

1) Objects <u>near</u> a radioactive source are <u>irradiated</u> by it. This simply means they're <u>exposed</u> to it (we're <u>always</u> being irradiated by <u>background radiation</u> sources).

2) <u>Irradiating</u> something does <u>not</u> make it <u>radioactive</u>.

3) Keeping sources in <u>lead-lined boxes</u>, standing behind <u>barriers</u> or being in a <u>different room</u> and using <u>remote-controlled arms</u> when working with radioactive sources are all ways of reducing <u>irradiation</u>.

See page 83 for the dangers of being exposed to radiation.

Your exposure to background radiation depends on where you live...

Background radiation comes from many sources, from <u>food and drink</u> to <u>cosmic rays</u>, but mostly it comes from the <u>ground</u> and is given out by certain rocks, like granite. That's why some parts of the UK have higher levels of <u>background radiation</u> than others. Areas like Cornwall and Devon, where there's lots of granite, have higher background radiation levels than is average for the UK. But they do have lovely beaches.

Contamination

Radioactive contamination is pretty dangerous — it comes about from touching and handling radioactive substances. Fortunately, there are some simple steps you can take to keep yourself safe from contamination.

Contamination is Radioactive Particles **Getting onto Objects**

1) If unwanted radioactive atoms get onto or into an object, the object is said to be contaminated. E.g. if you touch a radioactive source without wearing gloves, your hands would be contaminated.

2) These contaminating atoms might then decay, releasing radiation which could cause you harm.

3) Contamination is especially dangerous because radioactive particles could get inside your body.

4) Gloves and tongs should be used when handling sources, to avoid particles getting stuck to your skin or under your nails. Some industrial workers wear protective suits to stop them breathing in particles.

Exposure to Some Sources can be **More Harmful** than to Others

Contamination or irradiation can cause different amounts of harm, based on the radiation type.

1) Outside the body, beta and gamma sources are the most dangerous. This is because beta and gamma can penetrate the body and get to delicate organs. Alpha is less dangerous because it can't penetrate the skin and is easily blocked by a small air gap (p.75). High levels of irradiation from all sources are dangerous, but especially from ones that emit beta and gamma.

2) Inside the body, alpha sources are the most dangerous, because they do all their damage in a very localised area. So contamination, rather than irradiation, is the major concern when working with alpha sources. Beta sources are less damaging inside the body, as radiation is absorbed over a wider area, and some passes out of the body altogether. Gamma sources are the least dangerous inside the body, as they mostly pass straight out — they have the lowest ionising power, p.75.

The more we understand how different types of radiation affects our bodies, the better we can protect ourselves when using them. This is why it's so important that research about this is published. The data is peer-reviewed (see page 1) and can quickly become accepted, leading to many improvements in our use of radioactive sources.

Alpha sources are the most dangerous inside the body...

Alpha sources are the most ionising (see page 75), so if they get into the body, they can wreak havoc. Beta and gamma sources however are the most dangerous outside the body. This is because they are more penetrating than alpha sources (see page 75), so can get through the skin and cause damage to cells.

Uses and Risks of Radiation

Radiation can be pretty useful. We use it in our homes, in industry and in medicine. But it's not without its dangers. Using radiation is all about reducing the risks whilst still keeping the benefits.

Gamma Sources are Usually Used in Medical Tracers

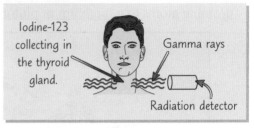

Iodine-123 collecting in the thyroid gland.

Gamma rays

Radiation detector

1) Certain radioactive isotopes can be injected into people (or they can just swallow them) and their progress around the body can be followed using an external detector. A computer converts the reading to a display showing where the strongest reading is coming from.

2) One example is the use of iodine-123, which is absorbed by the thyroid gland (just like normal iodine-127), but it gives out radiation which can be detected to indicate whether the thyroid gland is taking in iodine as it should.

3) Isotopes which are taken into the body like this are usually gamma emitters (never alpha emitters), so that the radiation passes out of the body without causing much ionisation. They should have a short half-life so that the radioactivity inside the patient quickly disappears.

Radiotherapy — Treating Cancer with Radiation

1) Since high doses of ionising radiation will kill all living cells (see below) it can be used to treat cancers.

2) Gamma rays are directed carefully and at just the right dosage to kill the cancer cells without damaging too many normal cells. Radiation-emitting implants (usually beta-emitters) can also be put next to or inside tumours.

3) However, a fair bit of damage is inevitably done to normal cells, which makes the patient feel very ill. But if the cancer is successfully killed off in the end, then it's worth it.

There are Risks to Using Radiation

1) Radiation can enter living cells and ionise atoms and molecules within them. This can lead to tissue damage.

2) Lower doses tend to cause minor damage without killing the cells. This can give rise to mutant cells which divide uncontrollably. This is cancer.

3) Higher doses tend to kill cells completely, causing radiation sickness (leading to vomiting, tiredness and hair loss) if a lot of cells all get blasted at once.

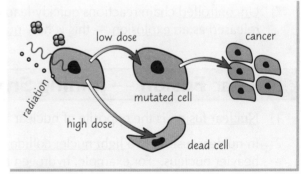

low dose — cancer

radiation

mutated cell

high dose

dead cell

- For every situation, it's worth considering both the benefits and risks of using radioactive materials.
- For example, tracers can be used to diagnose life-threatening conditions, while the risk of cancer from one use of a tracer is very small.
- Whilst prolonged exposure to radiation poses future risks (see p.140 for comparing the risks from different medical procedures) and causes many side effects, many people with cancer choose to have radiotherapy as it may get rid of their cancer entirely. For them, the benefits outweigh the risks.
- Perceived risk is how risky a person thinks something is. It's not the same as the actual risk of a procedure and the perceived risk can vary from person to person. See page 4 for more on this.

Nuclear Fission and Fusion

Splitting up atoms releases lots of useful energy, but it can have explosive consequences.

Nuclear Fission — Splitting a Large, Unstable Nucleus

Nuclear fission is a type of nuclear reaction that is used to release energy from large and unstable atoms (e.g. uranium or plutonium) by splitting them into smaller atoms.

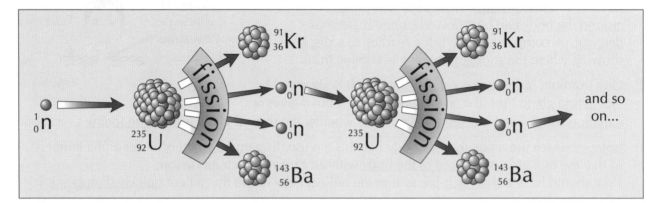

1) Spontaneous (unforced) fission rarely happens. Usually, the nucleus has to absorb a neutron before it will split.

You may have to draw or complete a diagram to show a chain reaction in the exam, so make sure you're happy with fission reactions.

2) When the atom splits it forms two new lighter elements that are roughly the same size (and that have some energy in their kinetic energy stores).

3) Two or three neutrons are also released when an atom splits. If any of these neutrons are moving slow enough to be absorbed by another nucleus, they can cause more fission to occur. This is a chain reaction.

4) The energy not transferred to the kinetic energy stores of the products is carried away by gamma rays.

5) The energy carried away by the gamma rays, and in the kinetic energy stores of the remaining free neutrons and other decay products, can be used to heat water, making steam to turn turbines and generators.

6) The amount of energy produced by fission in a nuclear reactor is controlled by changing how quickly the chain reaction can occur. This is done using control rods which, when lowered and raised inside a nuclear reactor, absorb neutrons. This slows down the chain reaction and controls the amount of energy released.

7) Uncontrolled chain reactions quickly lead to lots of energy being released as an explosion — this is how nuclear weapons work.

Nuclear Fusion — Joining Small Nuclei

1) Nuclear fusion is the opposite of nuclear fission.

2) In nuclear fusion, two light nuclei collide at high speed and join (fuse) to create a larger, heavier nucleus. For example, hydrogen nuclei can fuse to produce a helium nucleus.

3) This heavier nucleus does not have as much mass as the two separate, light nuclei did. Some of the mass of the lighter nuclei is converted to energy (don't panic, you don't need to know how). This energy is then released as radiation.

4) Fusion releases a lot of energy (more than fission for a given mass of fuel).

5) So far, scientists haven't found a way of using fusion to generate energy for us to use. The temperatures and pressures needed for fusion are so high that fusion reactors are really hard and expensive to build.

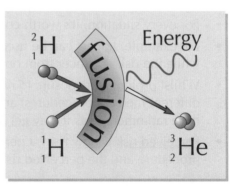

Warm-Up & Exam Questions

Radiation has pros and cons, and one of its downsides is that you need to know about it for your exam. So have a go at the practice exam questions below to see if you're radiating radiation knowledge.

Warm-Up Questions

1) Give three sources of background radiation.
2) What is the difference between radioactive contamination and irradiation?
3) Describe two precautions that should be taken when working with radioactive sources in a lab.
4) Why is nuclear radiation dangerous?
5) Why is contamination by an alpha source more dangerous to humans than irradiation by an alpha source?

Exam Questions

1 Nuclear radiation can have harmful effects on the human body. *(Grade 4-6)*

1.1 What is likely to happen to cells if they receive a high dose of nuclear radiation?

[1 mark]

1.2 Scientists have carried out research into the dangers of nuclear radiation.
Suggest **one** reason why it is important that the data from research is peer-reviewed.

[1 mark]

2 Nuclear reactors often use uranium-235. *(Grade 6-7)*

2.1 Describe how a chain reaction is set up in a nuclear reactor.

[3 marks]

2.2 In a nuclear reactor, control rods are used to control the reaction.
What could happen if the nuclear reaction is not controlled?

[1 mark]

3 Iodine-123 is a radioactive isotope, commonly used as a tracer in medicine. *(Grade 7-9)*

3.1 Describe how iodine-123 can be used to detect whether the thyroid gland is absorbing iodine as it normally should do.

[3 marks]

Table 1 shows the properties of three other radioisotopes.

Table 1

Radioisotope	Half-life	Type of emission
technetium-99m	6 hours	gamma
phosphorus-32	14 days	beta
cobalt-60	5 years	beta/gamma

3.2 State which of these would be best to use as a medical tracer. Explain your answer.

[3 marks]

Revision Summary for Topic 4

Well, that's the end of Topic 4 — hopefully it wasn't too painful. Time to see how much you've absorbed.
- Try these questions and tick off each one when you get it right.
- When you've done all the questions under a heading, and are completely happy with it, tick it off.

The Atomic Model (p.72-74) ☑

1) Briefly describe how the model of an atom has changed over time. ☑
2) Who provided evidence to suggest the existence of the neutron? ☑
3) Draw a sketch to show our currently accepted model of the atom. ☑
4) True or false? Atoms have no overall charge. ☑
5) What happens to an atom if it loses one or more of its outer electrons? ☑
6) Which number defines what element an atom is: the atomic number or the mass number? ☑
7) What is the atomic number of an atom? ☑
8) True or false? Isotopes have different mass numbers. ☑

Nuclear Decay and Half-life (p.75-78) ☑

9) What is radioactive decay? ☑
10) Name four things that may be emitted during radioactive decay. ☑
11) For alpha, beta and gamma radiation, give: a) their ionising power,   b) their range in air. ☑
12) Explain why alpha radiation could not be used to check the thickness of metal sheets. ☑
13) How could you represent alpha radiation in nuclear equations? ☑
14) What type of nuclear decay doesn't change the mass or charge of the nucleus? ☑
15) Name a piece of equipment that could be used to measure radiation. ☑
16) What is the activity of a source? What are its units? ☑
17) Explain the dangers of a radioactive source with a long half-life. ☑
18) Explain how you would find the half-life of a source, given a graph of its activity over time. ☑

Dangers and Uses of Radiation (p.81-83) ☑

19) Define radiation dose. ☑
20) State two aspects of your lifestyle that can affect your radiation dose. ☑
21) Define irradiation and contamination. ☑
22) Compare the hazards of being irradiated by an alpha source and a gamma source. ☑
23) Give two ways that radiation is used in medicine. ☑
24) Describe some of the risks involved with using radiation. ☑

Fission and Fusion (p.84) ☑

25) Define fission and fusion. ☑
26) True or false? Fission is usually spontaneous. ☑
27) Describe what a chain reaction is. ☑
28) Explain the difference between nuclear fission and nuclear fusion. ☑

Contact and Non-Contact Forces

Just like sports, forces are either <u>contact</u> or <u>non-contact</u> and involve lots of <u>interaction</u>.

Vectors Have **Magnitude** and **Direction**

1) Force is a <u>vector quantity</u> — vector quantities have a <u>magnitude</u> and a <u>direction</u>.

2) Lots of <u>physical quantities</u> are vector quantities:

> <u>Vector quantities</u>: force, velocity, displacement, acceleration, momentum, etc.

3) Some physical quantities <u>only</u> have magnitude and <u>no direction</u>.
   These are called <u>scalar quantities</u>:

> <u>Scalar quantities</u>: speed, distance, mass, temperature, time, etc.

4) Vectors are usually represented by an <u>arrow</u> — the <u>length</u> of the arrow shows the <u>magnitude</u>,
   and the <u>direction</u> of the arrow shows the <u>direction of the quantity</u>.

> <u>Velocity</u> is a <u>vector</u>, but <u>speed</u> is a <u>scalar</u> quantity.
> Both bikes are travelling at the same <u>speed</u>, v
> (the <u>length</u> of each arrow is the same).
> They have <u>different velocities</u> because
> they are travelling in different <u>directions</u>.

Forces Can be **Contact** or **Non-Contact**

1) A <u>force</u> is a <u>push</u> or a <u>pull</u> on an object that is caused by it <u>interacting</u> with something.

2) All forces are either <u>contact</u> or <u>non-contact</u> forces.

3) When <u>two objects</u> have to be <u>touching</u> for a force to act, that force is called a <u>contact force</u>.

> E.g. friction, air resistance, tension in ropes, normal contact force, etc.

4) If the objects <u>do not need to be touching</u> for the force to act, the force is a <u>non-contact force</u>.

> E.g. magnetic force, gravitational force, electrostatic force, etc.

5) When two objects <u>interact</u>, there is a <u>force</u> produced on <u>both</u> objects.
   An <u>interaction pair</u> is a pair of forces that are <u>equal</u> and <u>opposite</u> and act on two <u>interacting</u> objects.
   (This is basically Newton's Third Law — see p.112.)

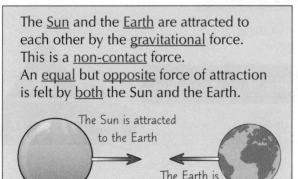

The <u>Sun</u> and the <u>Earth</u> are attracted to each other by the <u>gravitational</u> force.
This is a <u>non-contact</u> force.
An <u>equal</u> but <u>opposite</u> force of attraction is felt by <u>both</u> the Sun and the Earth.

The Sun is attracted to the Earth

The Earth is attracted to the Sun

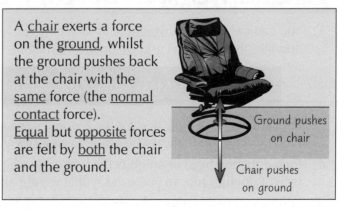

A <u>chair</u> exerts a force on the <u>ground</u>, whilst the ground pushes back at the chair with the <u>same</u> force (the <u>normal contact</u> force).
<u>Equal</u> but <u>opposite</u> forces are felt by <u>both</u> the chair and the ground.

Ground pushes on chair

Chair pushes on ground

Weight, Mass and Gravity

Gravity might seem like a rather heavy subject to tackle only a page into the topic, but you might end up finding it quite attractive. It's pretty important stuff, so make sure you understand it.

Gravitational Force is the Force of Attraction Between Masses

Gravity attracts all masses, but you only notice it when one of the masses is really really big, e.g. a planet. Anything near a planet or star is attracted to it very strongly.
This has two important effects:

> 1) On the surface of a planet, it makes all things fall towards the ground.

> 2) It gives everything a weight.

Weight and Mass are Not the Same

1) Mass is just the amount of 'stuff' in an object. For any given object this will have the same value anywhere in the universe.

2) Weight is the force acting on an object due to gravity (the pull of the gravitational force on the object). Close to Earth, this force is caused by the gravitational field around the Earth.

3) Gravitational field strength varies with location. It's stronger the closer you are to the mass causing the field, and stronger for larger masses.

4) The weight of an object depends on the strength of the gravitational field at the location of the object. This means that the weight of an object changes with its location.

5) For example, an object has the same mass whether it's on Earth or on the Moon — but its weight will be different. A 1 kg mass will weigh less on the Moon (about 1.6 N) than it does on Earth (about 9.8 N), simply because the gravitational field strength on the surface of the Moon is less.

6) Weight is a force measured in newtons. You can think of the force as acting from a single point on the object, called its centre of mass (a point at which you assume the whole mass is concentrated). For a uniform object (one that's the same density, p.64, throughout and is a regular shape), this will be at the centre of the object. ⟹

centre of mass

weight

7) Weight is measured using a calibrated spring balance (or newtonmeter).

8) Mass is not a force. It's measured in kilograms with a mass balance (an old-fashioned pair of balancing scales).

Mass and Weight are Directly Proportional

1) You can calculate the weight of an object if you know its mass (m) and the strength of the gravitational field that it is in (g):

> Weight (N) = Mass (kg) × Gravitational Field Strength (N/kg)

2) For Earth, $g \approx 9.8$ N/kg and for the Moon it's around 1.6 N/kg. Don't worry, you'll always be given a value of g to use in the exam.

3) Increasing the mass of an object increases its weight. If you double the mass, the weight doubles too, so you can say that weight and mass are directly proportional.

4) You can write this, using the direct proportionality symbol, as $W \propto m$.

Resultant Forces

When <u>multiple forces</u> act on an object, they can <u>add together</u> or <u>subtract</u> from each other until there's the equivalent of just <u>one</u> force acting in a <u>single direction</u>. This is the <u>resultant force</u>.

Free Body Diagrams Show All the Forces Acting on an Object

1) You need to be able to <u>describe</u> all the <u>forces</u> acting on an <u>isolated object</u> or a <u>system</u> (p.17) — i.e. <u>every</u> force <u>acting on</u> the object or system but <u>none</u> of the forces the object or system <u>exerts</u> on the rest of the world.

2) For example, a skydiver's <u>weight</u> acts on him, pulling him towards the ground, and <u>drag</u> (air resistance) also acts on him, in the <u>opposite direction</u> to his motion.

3) This can be shown using a <u>free body diagram</u> like the ones below.

4) The <u>sizes</u> of the arrows show the <u>relative magnitudes</u> of the forces and the <u>directions</u> show the directions of the forces acting on the object.

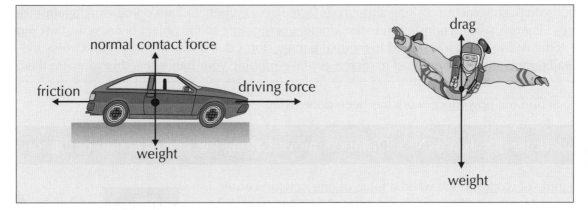

A Resultant Force is the Overall Force on a Point or Object

1) In most <u>real</u> situations there are at least <u>two forces</u> acting on an object along any direction.

2) If you have a <u>number of forces</u> acting at a single point, you can replace them with a <u>single force</u> (so long as the single force has the <u>same effect</u> as all the original forces together).

3) This single force is called the <u>resultant force</u>.
   For example, there is a <u>downward resultant force</u> acting on the <u>skydiver</u> above.

4) If the forces all act along the <u>same line</u> (they're all parallel), the <u>overall effect</u> is found by <u>adding</u> those going in the <u>same</u> direction and <u>subtracting</u> any going in the opposite direction.

EXAMPLE: **For the free body diagram shown on the right, calculate the resultant force acting on the van.**

1) Consider the <u>horizontal</u> and <u>vertical</u> directions <u>separately</u>.

   Vertical: 1500 − 1500 = 0 N
   Horizontal: 1200 − 1000 N = 200 N

2) State the <u>size</u> and <u>direction</u> of the <u>resultant</u> force.

   The resultant force is 200 N to the left.

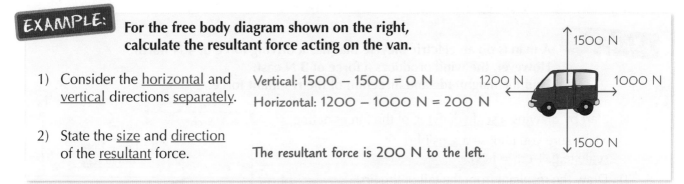

The resultant force — one force with the same result as many...

You'll most often encounter a <u>resultant force</u> as the <u>difference</u> between some kind of <u>driving force</u> and a <u>resistive force</u>, acting in <u>opposite directions</u> along the <u>same line</u>. For example, <u>weight</u> and <u>air resistance</u> for a <u>falling</u> object. But its not always so <u>straightforward</u>. Read on for some <u>techniques</u> for tackling them.

Resultant Forces

If A **Resultant** Force **Moves** An Object, **Work is Done**

> When a <u>force</u> moves an object through a <u>distance</u>,
> <u>ENERGY IS TRANSFERRED</u> and <u>WORK IS DONE</u> on the object.

1) To make something <u>move</u> (or <u>keep</u> it moving if there are <u>frictional forces</u>), a <u>force</u> must be applied.
2) The thing <u>applying the force</u> needs a <u>source</u> of <u>energy</u> (like <u>fuel</u> or <u>food</u>).
3) The force does '<u>work</u>' to <u>move</u> the object and <u>energy</u> is <u>transferred</u> from one <u>store</u> to another (p.18).
4) Whether energy is transferred '<u>usefully</u>' (e.g. <u>lifting a load</u>) or is '<u>wasted</u>' (p.23) you can still say that '<u>work is done</u>'. '<u>Work done</u>' and '<u>energy transferred</u>' are <u>the same thing</u>.

> When you push something along a <u>rough surface</u> (like a <u>carpet</u>) you are doing work <u>against frictional</u> <u>forces</u>. Energy is being <u>transferred</u> to the <u>kinetic energy store</u> of the <u>object</u> because it starts <u>moving</u>, but some is also being transferred to <u>thermal energy stores</u> due to the friction. This causes the overall <u>temperature</u> of the object to <u>increase</u>. (Like <u>rubbing your hands together</u> to warm them up.)

5) You can find out <u>how much</u> work has been done using:

> Work done (J) = Force (N) × Distance (moved along the line of action of the force) (m)

6) <u>One joule of work</u> is done when a <u>force of one newton</u> causes an object to move a <u>distance of one metre</u>. You need to be able to <u>convert</u> joules (J) to newton metres (Nm). <u>1 J = 1 Nm</u>.

$$W = Fs$$

$$\dfrac{W}{F \times s}$$

Use **Scale Drawings** to Find **Resultant Forces**

Working out resultant forces using scale diagrams isn't too tough. Here's what to do:

1) Draw all the <u>forces</u> acting on an object, to scale, '<u>tip-to-tail</u>'.
2) Then draw a <u>straight line</u> from the start of the <u>first force</u> to the <u>end</u> of the <u>last force</u> — this is the <u>resultant force</u>.
3) Measure the <u>length</u> of the <u>resultant force</u> on the diagram to find the <u>magnitude</u> and the <u>angle</u> to find the <u>direction</u> of the force.

EXAMPLE: **A man is on an electric bicycle that has a driving force of 4 N north.**
However, the wind produces a force of 3 N east.
Find the magnitude and direction of the resultant force.

1) Start by drawing a <u>scale drawing</u> of the forces acting.
2) Make sure you choose a <u>sensible</u> <u>scale</u> (e.g. 1 cm = 1 N).
3) Draw the <u>resultant</u> from the tail of the first arrow to the tip of the last arrow.
4) Measure the <u>length</u> of the resultant with a <u>ruler</u> and use the <u>scale</u> to find the force in N.
5) Use a <u>protractor</u> to measure the direction as a <u>bearing</u>. A bearing is an angle measured clockwise from north, given as a 3 digit number, e.g. 10° = 010°.

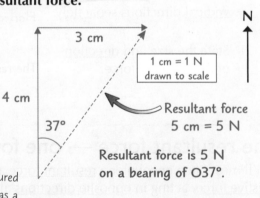

N

3 cm

1 cm = 1 N
drawn to scale

4 cm

Resultant force
5 cm = 5 N

37°

Resultant force is 5 N
on a bearing of 037°.

More on Forces

Scale diagrams are useful for more than just calculating resultant forces. You can also use them to check if forces are balanced and to split a force into component parts, as you're about to see...

An Object is in **Equilibrium** if the **Forces** on it are **Balanced**

1) If all of the forces acting on an object combine to give a resultant force of zero, the object is in equilibrium.

2) On a scale diagram, this means that the tip of the last force you draw should end where the tail of the first force you drew begins. E.g. for three forces, the scale diagram will form a triangle.

Make sure you draw the last force in the right direction. It's in the opposite direction to how you'd draw a resultant force.

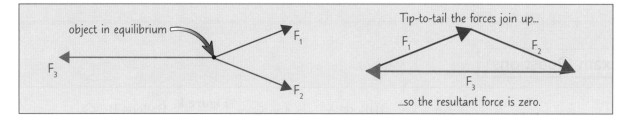

3) You might be given forces acting on an object and told to find a missing force, given that the object is in equilibrium. To do this, draw out the forces you do know (to scale and tip-to-tail), join the end of the last force to the start of the first force.

4) This line is the missing force so you can measure its size and direction.

You Can **Split** a Force into **Components**

1) Not all forces act horizontally or vertically — some act at awkward angles.

2) To make these easier to deal with, they can be split into two components at right angles to each other (usually horizontal and vertical).

3) Acting together, these components have the same effect as the single force.

4) You can resolve a force (split it into components) by drawing it on a scale grid.

EXAMPLE:

Use the grid below to resolve a 10 N force, acting at 53° above the horizontal, into horizontal and vertical components. Give your answers to 1 significant figure.

1) Begin by deciding on a scale for your grid. Here, we have 1 cm² squares, so an easy scale to work with would be if 1 cm = 1 N.

2) Next, draw your force to scale on the grid and at the right angle. Aim to have at least one end of the force arrow at the corner of a square on the grid.

3) Now draw a horizontal arrow from the bottom end of the force and a vertical arrow to the top end of the force to form a right angled triangle.

4) Measure the length of each arrow, and convert the lengths to N using your scale.

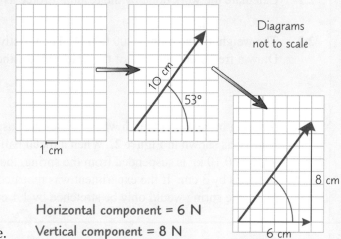

Horizontal component = 6 N

Vertical component = 8 N

With scale diagrams, it's important to keep things in proportion...

Scale diagrams are great for making calculations with forces, but you need to be careful. Make sure you keep your scale consistent (e.g. 1 square side = 1 N) and that you draw the forces in the correct direction.

Warm-Up & Exam Questions

Now you've learnt the basics of forces, it's time to act on your new knowledge.
Give these questions a whack and test how well you've forced those facts into your brain.

Warm-Up Questions

1) A tennis ball is dropped from a height.
   Name one contact force and one non-contact force that act on the ball as it falls.

2) Give an example of a vector quantity and a scalar quantity.

3) State the units of: a) gravitational field strength,   b) mass,   c) weight.

4) How can you tell if a set of forces are balanced using a scale diagram?

Exam Questions

1   **Figure 1** shows two hot air balloons, labelled with the forces acting on them.   `Grade 4-6`

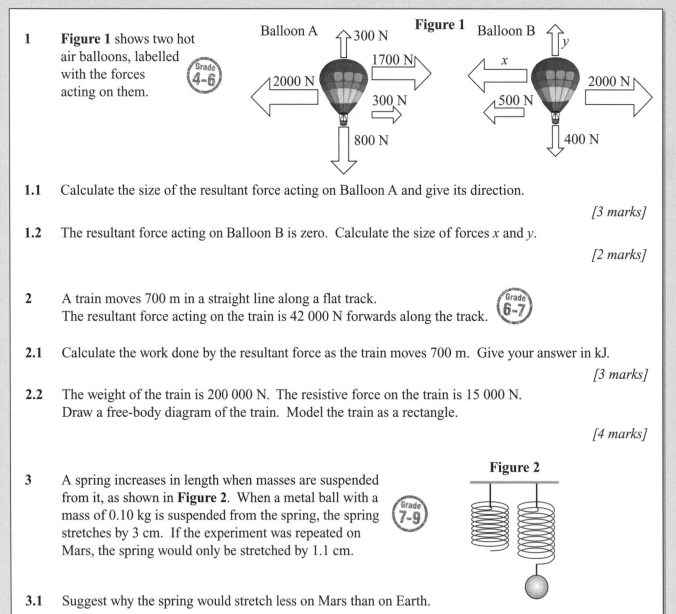

1.1   Calculate the size of the resultant force acting on Balloon A and give its direction.

[3 marks]

1.2   The resultant force acting on Balloon B is zero. Calculate the size of forces x and y.

[2 marks]

2   A train moves 700 m in a straight line along a flat track.   `Grade 6-7`
    The resultant force acting on the train is 42 000 N forwards along the track.

2.1   Calculate the work done by the resultant force as the train moves 700 m. Give your answer in kJ.

[3 marks]

2.2   The weight of the train is 200 000 N. The resistive force on the train is 15 000 N.
          Draw a free-body diagram of the train. Model the train as a rectangle.

[4 marks]

3   A spring increases in length when masses are suspended from it, as shown in **Figure 2**. When a metal ball with a mass of 0.10 kg is suspended from the spring, the spring stretches by 3 cm. If the experiment was repeated on Mars, the spring would only be stretched by 1.1 cm.   `Grade 7-9`

3.1   Suggest why the spring would stretch less on Mars than on Earth.

[3 marks]

3.2   The weight of the ball on Mars is 0.37 N. Calculate the value of g on Mars.

[3 marks]

Forces and Elasticity

Forces don't just make objects move, they can also make them change shape. Whether they change shape temporarily or permanently depends on the object and the forces applied.

Stretching, Compressing or Bending Transfers Energy

1) When you apply a force to an object you may cause it to stretch, compress or bend.

2) To do this, you need more than one force acting on the object — otherwise the object would simply move in the direction of the applied force, instead of changing shape.

3) Work is done when a force stretches or compresses an object and causes energy to be transferred to the elastic potential energy store of the object.

4) If it is elastically deformed (see below), ALL this energy is transferred to the object's elastic potential energy store (see p.19).

Elastic Deformation

1) An object has been elastically deformed if it can go back to its original shape and length after the force has been removed.

2) Objects that can be elastically deformed are called elastic objects (e.g. a spring).

Inelastic Deformation

1) An object has been inelastically deformed if it doesn't return to its original shape and length after the force has been removed.

Elastic objects are only elastic up to a certain point...

Remember the difference between elastic deformation and inelastic deformation. If an object has been elastically deformed, it will return to its original shape when you remove the force. If it's been inelastically deformed, its shape will have been changed permanently — for example, an over-stretched spring will stay stretched even after you remove the force.

Forces and Elasticity

Springs obey a really handy little equation that relates the force on them to their extension — for a while at least. Thankfully, you can plot a graph to see where this equation is valid.

Extension is **Directly Proportional** to **Force...**

If a spring is supported at the top and a weight is attached to the bottom, it stretches.

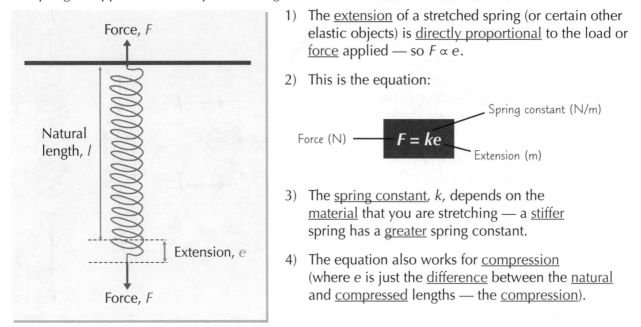

1) The extension of a stretched spring (or certain other elastic objects) is directly proportional to the load or force applied — so $F \propto e$.

2) This is the equation:

Spring constant (N/m)

Force (N) ——— $F = ke$

Extension (m)

3) The spring constant, k, depends on the material that you are stretching — a stiffer spring has a greater spring constant.

4) The equation also works for compression (where e is just the difference between the natural and compressed lengths — the compression).

...But this **Stops Working** when the **Force** is **Great Enough**

There's a limit to the amount of force you can apply to an object for the extension to keep on increasing proportionally.

1) The graph shows force against extension for an elastic object.

2) There is a maximum force above which the graph curves, showing that extension is no longer proportional to force.

3) This is known as the limit of proportionality and is shown on the graph at the point marked P.

4) You might see graphs with these axes the other way around — extension-force graphs. The graph still starts with a straight part, but starts to curve upwards once you go past the limit of proportionality, instead of downwards.

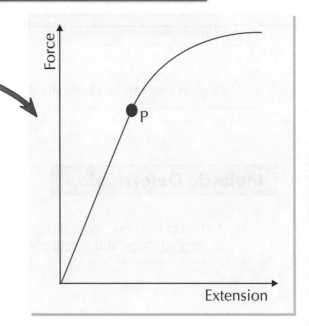

The spring constant is measured in N/m...

EXAM TIP

Be careful with units when doing calculations with springs. Your values for extension will usually be in centimetres or millimetres, but the spring constant is measured in newtons per metre. So convert the extension into metres before you do any calculations, or you'll get the wrong answer.

Investigating Springs

Oh look, here's another one of those <u>Required Practicals</u>... This one looks pretty straightforward, but there are a few ways this experiment can <u>stretch</u> you. <u>Read on</u>, so you won't be past your limits in the exam.

You Can **Investigate** the Link Between **Force** and **Extension**

1) Before you start, set up the <u>apparatus</u> as shown in the diagram. Make sure you have plenty of extra masses.

2) It's a good idea to measure the <u>mass</u> of each of your masses (with a mass balance) and calculate its <u>weight</u> (the <u>force</u> applied) using $W = mg$ (p.88) at this point. This'll mean you don't have to do a load of calculations in the middle of the experiment.

Before you launch into the investigation, you could do a quick <u>pilot experiment</u> to check your masses are an appropriate size for your investigation:

- Using an <u>identical spring</u> to the one you'll be testing, <u>load</u> it with <u>masses</u> one at a time up to a total of <u>five</u>. Measure the <u>extension each time</u> you add another mass.

- Work out the <u>increase</u> in the extension of the spring for <u>each</u> of your masses. If any of them cause a <u>bigger increase</u> in extension than the previous masses, you've gone past the spring's <u>limit of proportionality</u>. If this happens, you'll need to use <u>smaller masses</u>, or else you won't get enough measurements for your graph.

Method

1) Measure the <u>natural length</u> of the spring (when <u>no load</u> is applied) with a <u>millimetre ruler</u> clamped to the stand. Make sure you take the reading at eye level and add a <u>marker</u> (e.g. a thin strip of tape) to the <u>bottom</u> of the spring to make the reading more accurate.

2) Add a mass to the spring and allow the spring to come to <u>rest</u>. Record the mass and measure the new <u>length</u> of the spring. The <u>extension</u> is the change in length.

To check whether the deformation is elastic or inelastic, you can remove each mass temporarily and check to see if the spring goes back to the previous extension.

3) <u>Repeat</u> this process until you have enough measurements (no fewer than 6).

Extension is the change in length due to an applied force...

Make sure you know how to calculate the <u>extension of a spring</u>. It's <u>not</u> the just the <u>length</u> of the spring, it's the <u>difference</u> between the <u>stretched length</u> and the <u>original, unstretched length</u>. The extension when <u>no force</u> is acting on a spring should always be <u>zero</u> — unless the spring has been <u>inelastically deformed</u>.

Investigating Springs

Once you've collected all your <u>data</u>, you need to know what to do with it. Fortunately this page is all about how to use the <u>results</u> from the <u>practical</u> on the last page to work out things like the <u>spring constant</u>.

You Can **Plot** Your **Results** on a **Force-Extension Graph**

Once you've collected your results using the method on the last page, you can <u>plot</u> a <u>force-extension graph</u> of your results. It will only start to <u>curve</u> if you <u>exceed</u> the <u>limit of proportionality</u>, but don't worry if yours doesn't (as long as you've got the straight line bit).

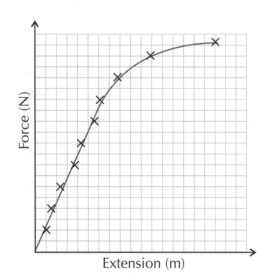

- When the line of best fit is a <u>straight line</u> it means there is a <u>linear</u> relationship between force and extension (they're <u>directly proportional</u>, see previous page). $F = ke$, so the <u>gradient</u> of the straight line is equal to k, the <u>spring constant</u>.

- When the line begins to <u>bend</u>, the relationship is now <u>non-linear</u> between force and extension — the spring <u>stretches more</u> for each unit increase in force.

You Can **Work Out Energy** Stored for **Linear** Relationships

1) As long as a spring is not stretched <u>past</u> its <u>limit of proportionality</u>, <u>work done</u> in stretching (or compressing) a spring can be found using:

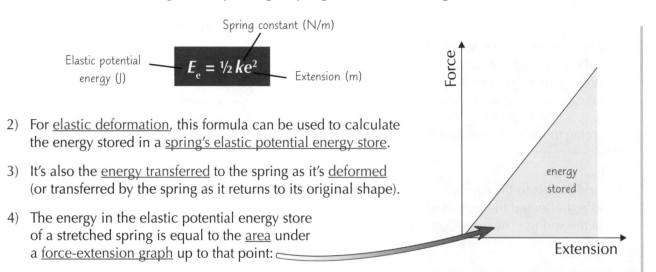

Spring constant (N/m)

Elastic potential energy (J) $E_e = \frac{1}{2} ke^2$ Extension (m)

2) For <u>elastic deformation</u>, this formula can be used to calculate the energy stored in a <u>spring's elastic potential energy store</u>.

3) It's also the <u>energy transferred</u> to the spring as it's <u>deformed</u> (or transferred by the spring as it returns to its original shape).

4) The energy in the elastic potential energy store of a stretched spring is equal to the <u>area</u> under a <u>force-extension graph</u> up to that point:

The force-extension graph curves at the limit of proportionality...

In reality, you may not always see the <u>curved part</u> in your force-extension graph for this experiment. This may be because you may not have added <u>enough masses</u> to your spring to reach, and go past, the <u>limit of proportionality</u>. But you can still use the <u>gradient</u> of your straight line to calculate the <u>spring constant</u>.

Moments

Forces can do a lot of nifty things when they act on objects, as I'm sure you've already seen. But there's more... This page is all about how they can make objects rotate. Give it a read, it'll only take a moment.

A **Moment** is the **Turning Effect** of a Force

A force, or several forces, can cause an object to rotate.

The turning effect of a force is called its moment.

The size of the moment of the force is given by:

Moment of a force (Nm) — $M = Fd$ — Distance (m) — the perpendicular distance from the pivot to the line of action of the force

Force (N)

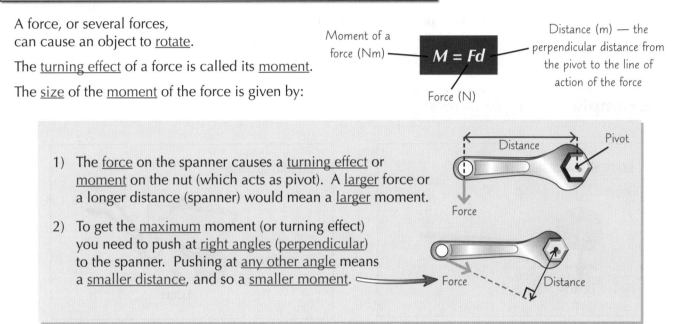

1) The force on the spanner causes a turning effect or moment on the nut (which acts as pivot). A larger force or a longer distance (spanner) would mean a larger moment.

2) To get the maximum moment (or turning effect) you need to push at right angles (perpendicular) to the spanner. Pushing at any other angle means a smaller distance, and so a smaller moment.

If the total anticlockwise moment equals the total clockwise moment about a pivot, the object is balanced and won't turn. You can use the equation above to find a missing force or distance in these situations.

EXAMPLE:

A 6 m long steel girder weighing 1000 N rests horizontally on a pole 1 m from one end. What is the tension in a supporting cable attached vertically to the other end?

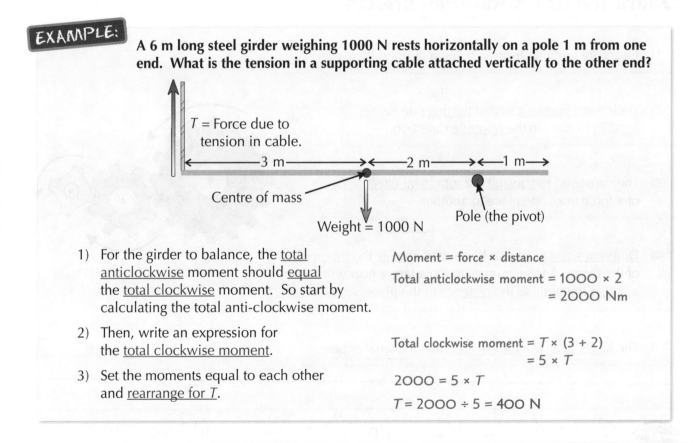

1) For the girder to balance, the total anticlockwise moment should equal the total clockwise moment. So start by calculating the total anti-clockwise moment.

   Moment = force × distance
   Total anticlockwise moment = 1000 × 2
   = 2000 Nm

2) Then, write an expression for the total clockwise moment.

   Total clockwise moment = T × (3 + 2)
   = 5 × T

3) Set the moments equal to each other and rearrange for T.

   2000 = 5 × T
   T = 2000 ÷ 5 = 400 N

Moment of a force = force × distance...

You'd better remember this equation. There may just be a moment in the exam where you need it...

Levers and Gears

<u>Moments</u> play a part in a lot of mechanical devices, from clocks to wheelbarrows. They can help you get the most out of a <u>small force</u>, and allow you to <u>transmit</u> forces across a distance.

Levers Make it Easier for us to Do Work

Levers <u>increase</u> the <u>distance</u> from the pivot at which the <u>force</u> is applied.
Since $M = Fd$ this means <u>less force</u> is needed to get the <u>same moment</u>.
This means levers make it <u>easier</u> to do <u>work</u>, e.g. <u>lift a load</u> or <u>turn a nut</u>.

Examples of Simple Levers

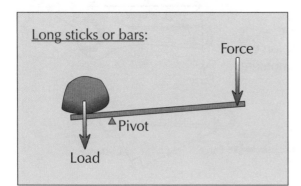

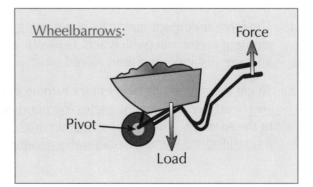

Gears Transmit Rotational Effects

1) <u>Gears</u> are circular discs with '<u>teeth</u>' around their edges.

2) Their teeth <u>interlock</u> so that <u>turning</u> one causes <u>another</u> to turn, in the <u>opposite</u> direction.

3) They are used to <u>transmit</u> the <u>rotational effect</u> of a <u>force</u> from one place to another.

4) <u>Different sized</u> gears can be used to <u>change</u> the <u>moment</u> of the force. A force transmitted to a <u>larger</u> gear will cause a <u>bigger</u> moment, as the <u>distance</u> to the pivot is greater.

5) The <u>larger gear</u> will <u>turn slower</u> than the smaller gear.

You need to be able to explain how gears and levers work...

<u>Gears</u> can be tricky to keep track of. Remember that, when you have a series of connected gears, the <u>direction</u> of rotation <u>changes</u> each time the rotational effect is transferred. Try drawing arrows to show <u>how</u> the gears <u>move</u>, so you can more easily follow how the <u>turning effect</u> is transmitted.

Warm-Up & Exam Questions

I can hear those gears whirring away in your brain from all that new information. So it's the perfect time to test what you know. Work through these questions then check to see if you got them right...

Warm-Up Questions

1) State the formula linking force, extension and spring constant.
2) What is meant by the limit of proportionality of a spring?
3) Give the equation used to calculate the moment of a force (around a pivot).
4) What are the units of the moment of a force?

Exam Questions

PRACTICAL

1 The teacher shows his students an experiment to show how a spring extends when masses are hung from it. He hangs a number of 90 g masses from a 50 g hook attached to the base of the spring. He records the extension of the spring and the total weight of the masses and hook each time he adds a mass to the bottom of the spring.

Grade 6-7

Figure 1

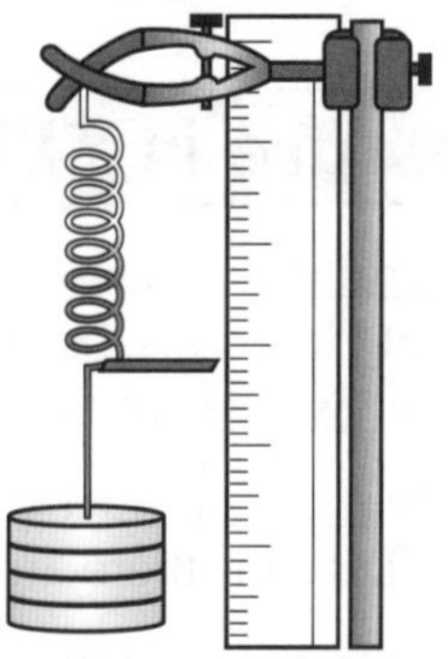

1.1 Give the independent variable in this experiment

[1 mark]

1.2 Give **one** control variable in this experiment.

[1 mark]

When a force of 4 N is applied to the spring, the spring extends elastically by 2.5 cm.

1.3 Calculate the spring constant of the spring.

[4 marks]

1.4 The teacher applies a 15 N force to the spring. When he removes the force, the spring is 7 cm long. The original length of the spring was 5 cm. Describe what has happened to the spring.

[1 mark]

2 Bolts are often secured using an Allen key, as shown in **Figure 2**. One end of the Allen key is put into the bolt and the other is turned to tighten the bolt.

Grade 6-7

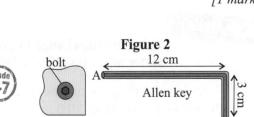

Figure 2

2.1 Calculate the moment on the bolt when end A of the Allen key is put into the bolt and a force of 15 N is applied to end B.

[3 marks]

2.2 Calculate the moment on the bolt when end B of the Allen key is put into the bolt and a force of 15 N is applied to end A.

[3 marks]

2.3 Which end of the Allen key (A or B) should be put into the bolt to make it easier to tighten? Explain your answer.

[2 marks]

Fluid Pressure

I'm sure you're familiar with <u>pressure</u>, what with exams coming up, but it's an important concept in physics too.

Pressure is the Force per Unit Area

1) <u>Fluids</u> are substances that can 'flow' because their particles are able to <u>move around</u>.   *A fluid is either a <u>liquid</u> or a <u>gas</u>.*

2) As these particles move around, they <u>collide</u> with surfaces and <u>other particles</u>.

3) Particles are light, but they still have a <u>mass</u> and exert a <u>force</u> on the object they collide with. <u>Pressure</u> is <u>force per unit area</u>, so this means the particles exert a <u>pressure</u>.

4) The <u>pressure</u> of a fluid means a <u>force</u> is exerted <u>normal</u> (at <u>right angles</u>) to any <u>surface</u> in contact with the fluid.

5) You can calculate the <u>pressure</u> at the <u>surface</u> of a fluid by using:

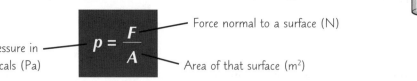

Force normal to a surface (N)

Pressure in pascals (Pa)

$$p = \frac{F}{A}$$

Area of that surface (m²)

Force

Force ← → Force

Force

Pressure in a Liquid Depends on Depth and Density

1) <u>Density</u> is a measure of the '<u>compactness</u>' of a substance, i.e. how <u>close together</u> the particles in a substance are (p.64). For a given <u>liquid</u>, the <u>density</u> is <u>uniform</u> (the <u>same</u> <u>everywhere</u>) and it <u>doesn't vary</u> with <u>shape</u> or <u>size</u>. The density of a gas can vary though.

2) The <u>more dense</u> a given liquid is, the <u>more particles</u> it has in a certain space. This means there are more particles that are able to <u>collide</u> so the <u>pressure is higher</u>.

3) As the <u>depth</u> of the liquid increases, the number of particles <u>above</u> that point increases. The weight of these particles adds to the pressure felt at that point, so liquid pressure increases with depth.

4) You can calculate the <u>pressure</u> at a certain <u>depth</u> due to the <u>column</u> of liquid <u>above</u> using:

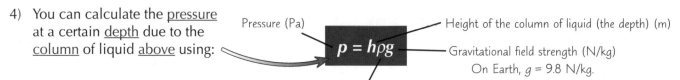

Pressure (Pa)

$$p = h\rho g$$

Height of the column of liquid (the depth) (m)

Gravitational field strength (N/kg) On Earth, g = 9.8 N/kg.

Density of the liquid (kg/m³) (the symbol is the Greek letter 'rho')

EXAMPLE:

Calculate the change in pressure between a point 25 m below the surface of water and a point 45 m below the surface. The density of water is 1000 kg/m³.

1) Calculate the <u>pressure</u> caused by the water at a depth of <u>20.0 m</u>.

$p = h\rho g$ = 25 × 1000 × 9.8
   = 245 000 Pa

2) Do the same for a depth of <u>40.0 m</u>.

$p = h\rho g$ = 45 × 1000 × 9.8
   = 441 000 Pa

Check your answer makes sense (you can't get negative pressure).

3) <u>Take away</u> the pressure at <u>20.0 m</u> from the pressure at <u>40.0 m</u>.

441 000 − 245 000 = 196 000 Pa
        (or 196 kPa)

You could write the answer above in <u>standard form</u> — it's a way of writing <u>very big</u> or <u>very small</u> numbers. Numbers in standard form always look like this:

A is always a number between 1 and 10.

A × 10ⁿ

n is the number of places the decimal point would move if you wrote the number out fully. It's negative for numbers less than 1, and positive for numbers greater than 1.

So 200 000 Pa would be written as 2×10^5 Pa in standard form.

Upthrust

This page is all about <u>upthrust</u> — a force that determines whether an object will <u>sink</u> or <u>float</u>. So take some time to read up on it now, so you're not <u>submerged</u> in revision at the last minute.

Objects in Fluids Experience Upthrust

1) When an object is submerged <u>in</u> a fluid (either partially or completely), the <u>pressure</u> of the fluid exerts a <u>force</u> on it from <u>every direction</u>.

2) Pressure <u>increases with depth</u>, so the force exerted on the <u>bottom</u> of the object is <u>larger than</u> the force acting on the <u>top</u> of the object.

Pressure

Pressure

Pineapple displaces this much water.

Upthrust is equal to the weight of this amount of water.

3) This causes a <u>resultant force</u> (p.89) upwards, known as <u>upthrust</u>.

4) The upthrust is <u>equal</u> to the <u>weight</u> of fluid that has been <u>displaced</u> (pushed out of the way) by the object. E.g. the upthrust on a pineapple in water is equal to the <u>weight</u> of a <u>pineapple-shaped volume</u> of water.

An Object Floats if its Weight = Upthrust

1) If the <u>upthrust</u> on an object is <u>equal to</u> the object's <u>weight</u>, then the forces <u>balance</u> and the object <u>floats</u>.

2) If an object's <u>weight</u> is <u>more than</u> the <u>upthrust</u>, the object <u>sinks</u>.

3) This means that whether or not an object will float depends on its <u>density</u>.

4) An object that is <u>less dense</u> than the fluid it is placed in <u>weighs less</u> than the <u>equivalent volume</u> of fluid. This means it <u>displaces</u> a <u>volume</u> of fluid that is <u>equal to its weight</u> before it is <u>completely submerged</u>.

5) At this point, the object's weight is <u>equal</u> to the upthrust, so the object <u>floats</u>.

6) An object that is <u>denser</u> than the fluid it is placed in is <u>unable</u> to displace enough fluid to equal its weight. This means that its weight is always <u>larger</u> than the upthrust, so it <u>sinks</u>.

This much water weighs the <u>same</u> as the whole apple (because the apple is <u>less dense</u> than water).

The apple has displaced a volume of water <u>equal</u> to its weight so it floats.

This much water weighs <u>less</u> than a potato (because the potato is <u>denser</u> than water).

The potato can <u>never</u> displace a volume of water equal to its weight so it sinks.

Submarines make use of <u>upthrust</u>. To <u>sink</u>, large tanks are <u>filled with water</u> to increase the <u>weight</u> of the submarine so that it is <u>more than</u> the upthrust. To rise to the surface, the tanks are filled with <u>compressed air</u> to reduce the weight so that it's <u>less than</u> the upthrust.

Atmospheric Pressure

Atmospheric pressure is exactly what it sounds like — it's the pressure exerted by the air around you. It's acting on you at all times, and varies with your height above sea-level.

The Atmosphere is the Air Surrounding the Earth

1) The atmosphere is a layer of air that surrounds Earth. It is thin compared to the size of the Earth.

2) Atmospheric pressure is created on a surface by air molecules colliding with the surface.

Atmospheric Pressure Decreases with Height

1) As the altitude (height above Earth) increases, atmospheric pressure decreases.

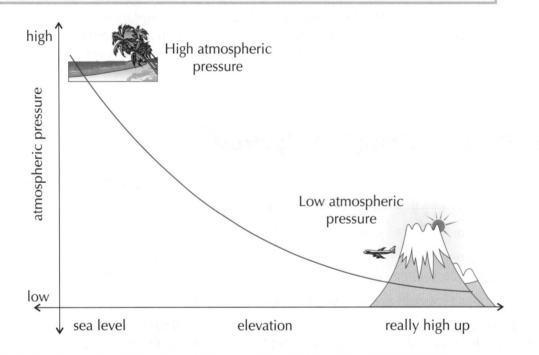

2) This is because as the altitude increases, the atmosphere gets less dense, so there are fewer air molecules that are able to collide with the surface.

3) There are also fewer air molecules above a surface as the height increases. This means that the weight of the air above it, which contributes to atmospheric pressure, decreases with altitude.

Atmospheric pressure decreases as height above ground increases...

...because the density of air decreases as height increases. This is why mountain climbers carry tanks of oxygen. The lower density of air means there are fewer air particles, and so less oxygen, which makes it harder to breath. So the mountaineers have to carry their own supply of oxygen with them.

Warm-Up & Exam Questions

Well, now you've gotten through that section, have a go at these questions, and see how much you've learnt. If you can master these, it should ease the pressure a bit for the exam.

Warm-Up Questions

1) Define the term pressure.
2) Other than density, name two variables the pressure due to a column of liquid depends on.
3) An object with a weight of 3.5 N is floating in a bucket of water.
   Give the size of the upthrust acting on the object.
4) How does atmospheric pressure vary with altitude?

Exam Questions

1 A student places a ball into a bucket of water
and it sinks to the bottom, as shown in **Figure 1**.
The water is 50 cm deep, and the ball has a diameter of 8 cm.

(Grade 6-7)

Figure 1

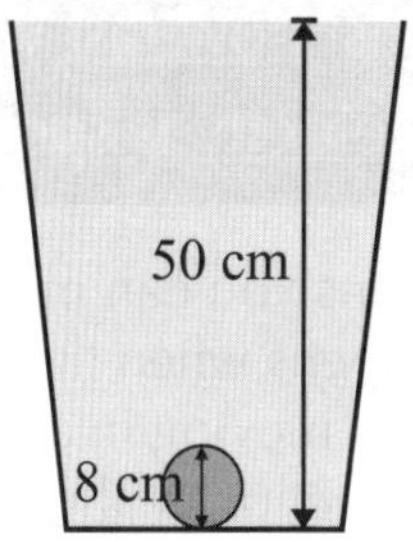

50 cm

8 cm

1.1 What conclusion can be made about the density of the ball?
Tick **one** box.

☐ It is more dense than water.

☐ It is less dense than water.

☐ It has the same density as water.

[1 mark]

1.2 Water has a density of 1000 kg/m³. The gravitational field strength is 9.8 N/kg.
Calculate the pressure due to the column of water above the ball.
Give your answer to 2 significant figures.
Use the correct equation from the Physics Equation Sheet on the inside back cover.

[4 marks]

2 **Figure 2** shows a simple hydraulic system containing a liquid. In a hydraulic system, pressure
is transmitted from one piston to another through the liquid. The liquid cannot be compressed.

(Grade 7-9)

Figure 2

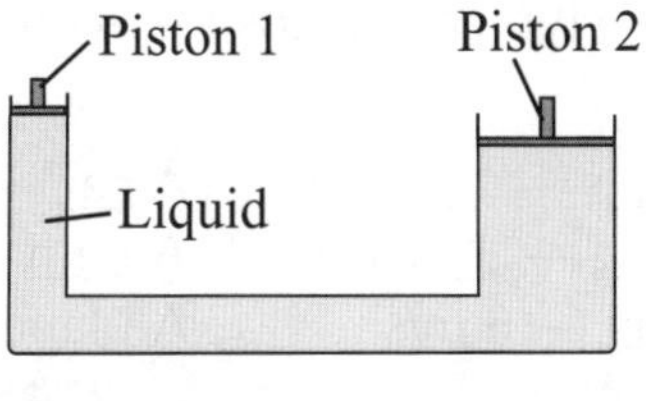

Piston 1   Piston 2

Liquid

2.1 A force of 175 N is applied to piston 1, which has a cross-sectional area of 0.25 m².
Calculate the pressure created at the first piston.

[2 marks]

2.2 Piston 2 has a cross-sectional area of 1.3 m².
Calculate the force acting on piston 2.

[3 marks]

Distance, Displacement, Speed and Velocity

There are a lot of very similar <u>variables</u> on this page, but they're <u>different</u> in some <u>very important</u> ways, so prepare to pay extra close attention. It's down to whether they're a <u>vector</u> or a <u>scalar</u> quantity.

Distance is Scalar, Displacement is a Vector

1) <u>Distance</u> is just <u>how far</u> an object has moved. It's a <u>scalar</u> quantity (p.87) so it doesn't involve <u>direction</u>.

2) Displacement is a <u>vector</u> quantity. It measures the distance and direction in a <u>straight line</u> from an object's <u>starting point</u> to its <u>finishing point</u> — e.g. the plane flew 5 metres <u>north</u>. The direction could be <u>relative to a point</u>, e.g. <u>towards the school</u>, or a <u>bearing</u> (a <u>three-digit angle from north</u>, e.g. <u>035˚</u>).

3) If you walk 5 m <u>north</u>, then 5 m <u>south</u>, your <u>displacement</u> is <u>0 m</u> but the <u>distance</u> travelled is <u>10 m</u>.

Speed and Velocity are Both How Fast You're Going

1) <u>Speed and velocity</u> both measure <u>how fast</u> you're going, but <u>speed</u> is a <u>scalar</u> and <u>velocity</u> is a <u>vector</u>:

<u>Speed</u> is just <u>how fast</u> you're going (e.g. 30 mph or 20 m/s) with no regard to the direction.
<u>Velocity</u> is speed in a given <u>direction</u>, e.g. 30 mph north or 20 m/s, 060°.

2) This means you can have objects travelling at a <u>constant speed</u> with a <u>changing velocity</u>. This happens when the object is <u>changing direction</u> whilst staying at the <u>same speed</u>. An object moving in a <u>circle</u> at a <u>constant speed</u> has a <u>constantly changing</u> velocity, as the direction is <u>always changing</u> (e.g. a <u>car</u> going around a <u>roundabout</u>).

3) If you want to <u>measure</u> the <u>speed</u> of an object that's moving with a <u>constant speed</u>, you should <u>time</u> how long it takes the object to travel a certain <u>distance</u>, e.g. using a <u>ruler</u> and a <u>stopwatch</u>. You can then <u>calculate</u> the object's <u>speed</u> from your measurements using this <u>formula</u>:

$$s = vt$$

distance travelled (m) = speed (m/s) × time (s)

4) Objects <u>rarely</u> travel at a <u>constant speed</u>. E.g. when you <u>walk</u>, <u>run</u> or travel in a <u>car</u>, your speed is <u>always changing</u>. For these cases, the formula above gives the <u>average</u> (<u>mean</u>) speed during that time.

You Need to Know Some Typical Everyday Speeds

1) Whilst every person, train, car etc. is <u>different</u>, there is usually a <u>typical speed</u> that each object travels at. <u>Remember</u> these typical speeds for everyday objects:

A person <u>walking</u> — <u>1.5 m/s</u>	A <u>car</u> — <u>25 m/s</u>
A person <u>running</u> — <u>3 m/s</u>	A <u>train</u> — <u>55 m/s</u>
A person <u>cycling</u> — <u>6 m/s</u>	A <u>plane</u> — <u>250 m/s</u>

2) Lots of different things can <u>affect</u> the speed something travels at. For example, the speed at which a person can <u>walk</u>, <u>run</u> or <u>cycle</u> depends on their <u>fitness</u>, their <u>age</u>, the <u>distance travelled</u> and the <u>terrain</u> (what kind of <u>land</u> they're moving over, e.g. roads, fields) as well as many other factors.

3) It's not only the speed of <u>objects</u> that varies. The speed of <u>sound</u> (<u>330 m/s</u> in <u>air</u>) <u>changes</u> depending on what the sound waves are <u>travelling</u> through, and the <u>speed of wind</u> is affected by many factors.

4) Wind speed can be affected by things like <u>temperature</u>, atmospheric <u>pressure</u> and if there are any large <u>buildings</u> or structures nearby (e.g. forests reduce the speed of the air travelling through them).

Acceleration

Acceleration is the rate of change of velocity. For cases of constant acceleration, there's a really useful equation you can use to calculate all sorts of variables of motion.

Acceleration is How Quickly You're Speeding Up

1) Acceleration is definitely not the same as velocity or speed.
2) Acceleration is the change in velocity in a certain amount of time.
3) You can find the average acceleration of an object using:

Acceleration (m/s²) —— $a = \dfrac{\Delta v}{t}$ —— Change in velocity (m/s)
—— Time (s)

EXAMPLE:

A cat accelerates at 2.5 m/s²
from 2.0 m/s to 6.0 m/s.
Find the time it takes to do this.

$t = \Delta v \div a$
$= (6.0 - 2.0) \div 2.5 = 1.6$ s

4) Deceleration is just negative acceleration (if something slows down, the change in velocity is negative).

You Need to be Able to Estimate Accelerations

You might have to estimate the acceleration (or deceleration) of an object.
To do this, you need the typical speeds from the previous page:

EXAMPLE: A car is travelling along a road, when it collides with a tree and comes to a stop. Estimate the deceleration of the car.

1) First, give a sensible speed for the car to be travelling at. The typical speed of a car is ~25 m/s.
2) Next, estimate how long it would take the car to stop. The car comes to a stop in ~1 s.
3) Put these numbers into the acceleration equation.

$a = \Delta v \div t$
$= (-25) \div 1$
$= -25$ m/s²

The ~ symbol just means it's an approximate value (or answer).

4) The question asked for the deceleration, so you can lose the minus sign (which shows the car is slowing down): So the deceleration is ~25 m/s²

Uniform Acceleration Means a Constant Acceleration

1) Constant acceleration is sometimes called uniform acceleration.
2) Acceleration due to gravity (g) is uniform for objects in free fall. It's roughly equal to 9.8 m/s² near the Earth's surface and has the same value as gravitational field strength (p.88).
3) You can use this equation for uniform acceleration:

Final velocity (m/s) —— $v^2 - u^2 = 2as$ —— Acceleration (m/s²)
—— Distance (m)
Initial velocity (m/s)

Initial velocity is just the starting velocity of the object.

EXAMPLE: A van travelling at 23 m/s starts decelerating uniformly at 2.0 m/s² as it heads towards a built-up area 112 m away. What will its speed be when it reaches the built-up area?

1) First, rearrange the equation so v^2 is on one side. $v^2 = u^2 + 2as$
2) Now put the numbers in — remember a is negative because it's a deceleration. $v^2 = 23^2 + (2 \times -2.0 \times 112)$
$= 81$
3) Finally, square root the whole thing. $v = \sqrt{81} = 9$ m/s

Distance-Time Graphs

It's time for some more exciting graphs. Distance-time graphs contain a lot of information, but they can look a bit complicated. Read on to get to grips with the rules of the graphs, and all will become clear.

You Can **Show Journeys** on **Distance-Time Graphs**

If an object moves in a straight line, its distance travelled can be plotted on a distance-time graph.

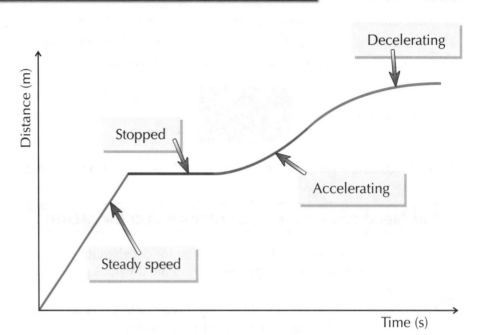

The **Shape of the Graph** Shows How the **Object** is **Moving**

1) Gradient = speed. (The steeper the graph, the faster the object is going.)
   This is because: speed = distance ÷ time = (change in vertical axis) ÷ (change in horizontal axis).

2) Flat sections are where the object's stationary — it's stopped.

3) Straight uphill sections mean it is travelling at a steady speed.

4) Curves represent acceleration or deceleration (p.105).

5) A steepening curve means the object's speeding up (increasing gradient).

6) A levelling off curve means it's slowing down (decreasing gradient).

7) If the object is changing speed (accelerating) you can find its speed at a point by finding the gradient of the tangent to the curve at that point, p.10.

 ## Read the axes of any graph you get given carefully...

Make sure you don't get confused between distance-time graphs and velocity-time graphs. They can look similar, but tell you different things and have different rules, as you're about to find out...

Velocity-Time Graphs

Even more <u>graphs</u>! Just like <u>distance-time graphs</u>, <u>velocity-time graphs</u> are a great way of <u>representing</u> <u>journeys</u>. There's a lot of <u>information</u> in them, so make sure you know how to get the most out of them.

You Can Also Show them on a **Velocity-Time Graph**

How an object's <u>velocity</u> changes as it travels can be plotted on a <u>velocity-time</u> graph.

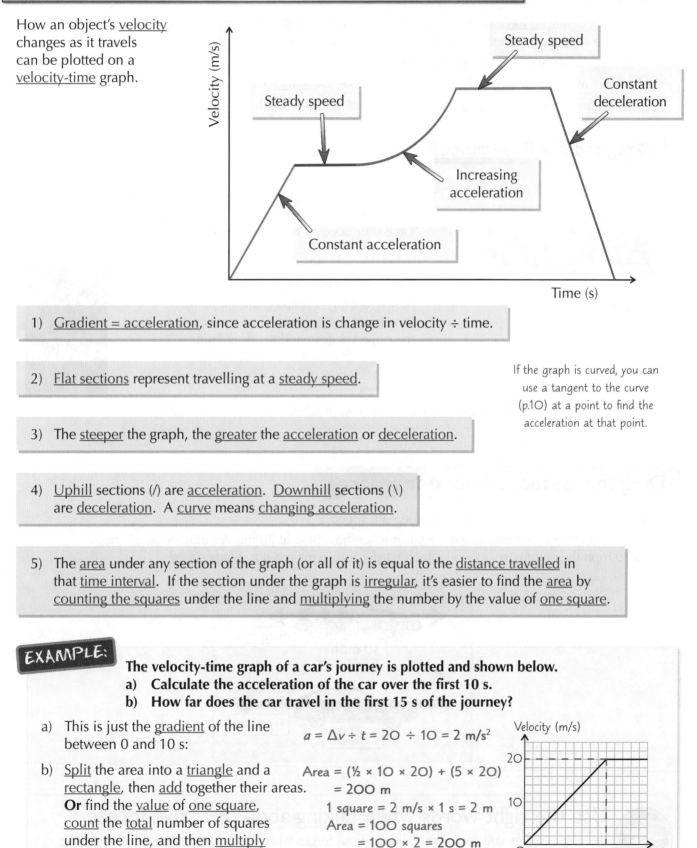

1) <u>Gradient = acceleration</u>, since acceleration is change in velocity ÷ time.

2) <u>Flat sections</u> represent travelling at a <u>steady speed</u>.

If the graph is curved, you can use a tangent to the curve (p.10) at a point to find the acceleration at that point.

3) The <u>steeper</u> the graph, the <u>greater</u> the <u>acceleration</u> or <u>deceleration</u>.

4) <u>Uphill</u> sections (/) are <u>acceleration</u>. <u>Downhill</u> sections (\) are <u>deceleration</u>. A <u>curve</u> means <u>changing acceleration</u>.

5) The <u>area</u> under any section of the graph (or all of it) is equal to the <u>distance travelled</u> in that <u>time interval</u>. If the section under the graph is <u>irregular</u>, it's easier to find the <u>area</u> by <u>counting the squares</u> under the line and <u>multiplying</u> the number by the value of <u>one square</u>.

EXAMPLE:

The velocity-time graph of a car's journey is plotted and shown below.
a) Calculate the acceleration of the car over the first 10 s.
b) How far does the car travel in the first 15 s of the journey?

a) This is just the <u>gradient</u> of the line between 0 and 10 s:

$a = \Delta v \div t = 20 \div 10 = 2 \text{ m/s}^2$

b) <u>Split</u> the area into a <u>triangle</u> and a <u>rectangle</u>, then <u>add</u> together their areas.
Or find the <u>value</u> of <u>one square</u>, <u>count</u> the <u>total</u> number of squares under the line, and then <u>multiply</u> these two values together.

Area = (½ × 10 × 20) + (5 × 20)
= 200 m
1 square = 2 m/s × 1 s = 2 m
Area = 100 squares
= 100 × 2 = 200 m

Topic 5 — Forces

Drag

Revision can be a bit of a <u>drag</u>, but hey, you're over halfway though the topic now.
No use <u>slowing down</u> now — however, there's quite a bit of that on this page.

Friction is Always There to **Slow Things Down**

1) If an object has <u>no force</u> propelling it along it will always <u>slow down and stop</u>
   because of <u>friction</u> (unless you're in space where there's nothing to rub against).

2) Friction always acts in the <u>opposite</u> direction to movement.

3) To travel at a <u>steady</u> speed, the driving force needs to <u>balance</u> the frictional forces.

4) You get friction between <u>two surfaces</u> in contact, or when an object passes <u>through a fluid</u> (<u>drag</u>).

Drag and **Air Resistance**

Air flows easily
over a streamlined car.

1) <u>Drag</u> is the <u>resistance</u> you get in a <u>fluid</u> (a gas or a liquid).
   <u>Air resistance</u> is a type of <u>drag</u> — it's the frictional
   force produced by the <u>air</u> acting on a <u>moving object</u>.

2) The most <u>important factor</u> by far in
   reducing drag is keeping the shape of
   the object <u>streamlined</u>. This is where
   the object is designed to allow fluid
   to <u>flow easily</u> across it, reducing drag.

3) Parachutes work in the <u>opposite</u> way — they want as much drag as they can get.

Drag **Increases** as **Speed Increases**

<u>Frictional forces</u> from fluids always <u>increase with speed</u>.

A car has <u>much more</u> friction to <u>work against</u> when travelling at <u>70 mph</u> compared to <u>30 mph</u>.
So at 70 mph the engine has to work <u>much harder</u> just to maintain a <u>steady speed</u>.

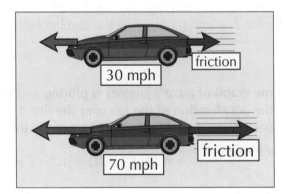

 ## Use the right words when talking about resistive forces...

EXAM TIP Be as <u>specific</u> as you can when talking about <u>forces</u> which act <u>against motion</u> (i.e. <u>resistive forces</u>). If an object is travelling through <u>air</u>, simply referring to the resistive force as 'drag' may not be enough to get you the marks — you'll need to specify the force is <u>air resistance</u>.

Terminal Velocity

If an object <u>falls</u> for long enough, it will reach its <u>terminal velocity</u>. It's all about <u>balance</u> between <u>weight</u> and <u>air resistance</u>. <u>Parachutes</u> work by <u>decreasing</u> your terminal velocity.

Objects **Falling** Through **Fluids** Reach a **Terminal Velocity**

1) When falling objects first <u>set off</u>, the force of gravity is <u>much more</u> than the <u>frictional force</u> slowing them down, so they accelerate.

2) As the <u>speed increases</u> the friction <u>builds up</u>.

3) This gradually <u>reduces</u> the <u>acceleration</u> until eventually the <u>frictional force</u> is <u>equal</u> to the <u>accelerating force</u> (so the <u>resultant force is zero</u>).

4) It will have reached its maximum speed or <u>terminal velocity</u> and will fall at a steady speed.

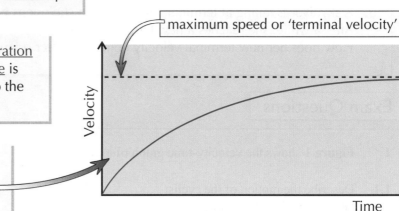

maximum speed or 'terminal velocity'

Velocity

Time

Terminal Velocity Depends on **Shape** and **Area**

1) The <u>accelerating force</u> acting on <u>all</u> falling objects is <u>gravity</u> and it would make them all fall at the <u>same</u> rate, if it wasn't for <u>air resistance</u>.

2) This means that on the Moon, where there's <u>no air</u>, hammers and feathers dropped simultaneously will hit the ground <u>together</u>.

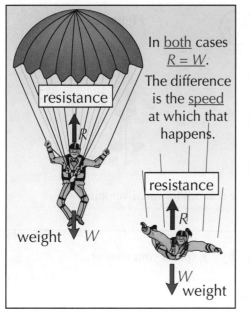

In <u>both</u> cases <u>R = W</u>. The difference is the <u>speed</u> at which that happens.

resistance

R

weight W

resistance

R

W
weight

3) However, on Earth, <u>air resistance</u> causes things to fall at <u>different</u> speeds, and the <u>terminal velocity</u> of any object is determined by its <u>drag</u> in <u>comparison</u> to its <u>weight</u>.

4) The frictional force depends on its <u>shape and area</u>. An important example is the human <u>skydiver</u>.

5) Without his parachute open he has quite a <u>small</u> area and a force of "<u>W = mg</u>" pulling him down.

6) He reaches a <u>terminal velocity</u> of about <u>120 mph</u>.

7) But with the parachute <u>open</u>, there's much more <u>air resistance</u> (at any given speed) and still only the same force "<u>W = mg</u>" pulling him down.

8) This means his <u>terminal velocity</u> comes right down to about <u>15 mph</u>, which is a <u>safe speed</u> to hit the ground at.

Warm-Up & Exam Questions

Slow down, it's not time to move on to the next section just yet. First it's time to check that all the stuff you've just read is still running around your brain. Dive into these questions.

Warm-Up Questions

1) What is the difference between speed and velocity?
2) Suggest the typical speeds of: a) a person running,   b) a train,   c) a plane.
3) How is acceleration shown on a distance-time graph?
4) Describe the shape of the line on a velocity-time graph for an object travelling at a steady speed.
5) In general, how does the air resistance acting on a car change as its speed increases?
6) A skydiver is falling at terminal velocity when she opens her parachute.
   After a while, she reaches a new terminal velocity.
   How does her new terminal velocity compare to her original terminal velocity?

Exam Questions

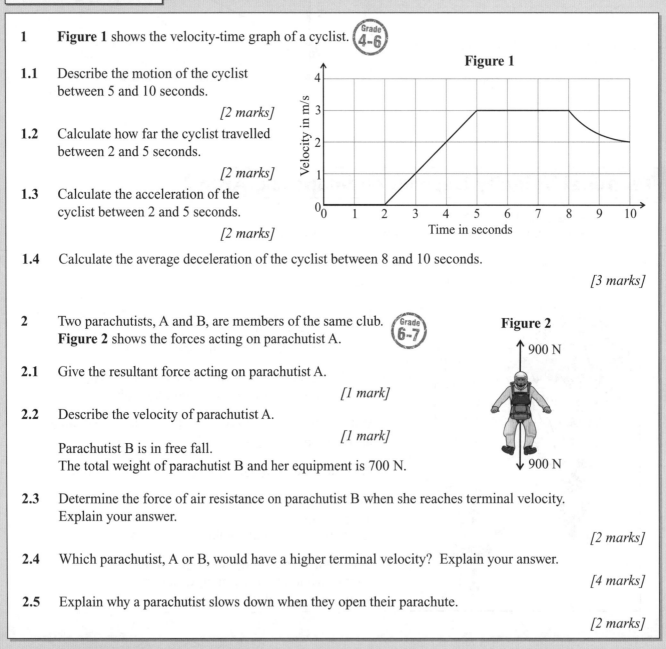

1 **Figure 1** shows the velocity-time graph of a cyclist. (Grade 4-6)

Figure 1

1.1 Describe the motion of the cyclist between 5 and 10 seconds.

[2 marks]

1.2 Calculate how far the cyclist travelled between 2 and 5 seconds.

[2 marks]

1.3 Calculate the acceleration of the cyclist between 2 and 5 seconds.

[2 marks]

1.4 Calculate the average deceleration of the cyclist between 8 and 10 seconds.

[3 marks]

2 Two parachutists, A and B, are members of the same club. (Grade 6-7)
Figure 2 shows the forces acting on parachutist A.

Figure 2

900 N

900 N

2.1 Give the resultant force acting on parachutist A.

[1 mark]

2.2 Describe the velocity of parachutist A.

[1 mark]

Parachutist B is in free fall.
The total weight of parachutist B and her equipment is 700 N.

2.3 Determine the force of air resistance on parachutist B when she reaches terminal velocity.
Explain your answer.

[2 marks]

2.4 Which parachutist, A or B, would have a higher terminal velocity? Explain your answer.

[4 marks]

2.5 Explain why a parachutist slows down when they open their parachute.

[2 marks]

Newton's First and Second Laws

Way back in the 1660s, some clever chap named <u>Isaac Newton</u> worked out some <u>Laws of Motion</u>...

A **Force** is Needed to **Change Motion**

This may seem simple, but it's important. <u>Newton's First Law</u> says that a resultant force (p.89) is needed to make something <u>start moving</u>, <u>speed up</u> or <u>slow down</u>:

> If the resultant force on a <u>stationary</u> object is <u>zero</u>, the object will <u>remain stationary</u>. If the <u>resultant force</u> on a moving object is <u>zero</u>, it'll just carry on moving at the <u>same velocity</u> (same speed <u>and</u> direction).

So, when a train or car or bus or anything else is <u>moving</u> at a <u>constant velocity</u>, the resistive and driving <u>forces</u> on it must all be <u>balanced</u>. The velocity will only change if there's a <u>non-zero</u> resultant force acting on the object.

1) A non-zero <u>resultant</u> force will always produce <u>acceleration</u> (or deceleration) in the <u>direction of the force</u>.

2) This "<u>acceleration</u>" can take <u>five</u> different forms: <u>starting</u>, <u>stopping</u>, <u>speeding up</u>, <u>slowing down</u> and <u>changing direction</u>.

3) On a free body diagram, the <u>arrows</u> will be <u>unequal</u>.

Acceleration is **Proportional** to the **Resultant Force**

1) The <u>larger</u> the <u>resultant force</u> acting on an object, the <u>more</u> the object accelerates — the force and the acceleration are <u>directly proportional</u>. You can write this as $F \propto a$.

2) Acceleration is also <u>inversely proportional</u> to the <u>mass</u> of the object — so an object with a <u>larger</u> mass will accelerate <u>less</u> than one with a smaller mass (for a <u>fixed resultant force</u>).

3) There's an incredibly <u>useful formula</u> that describes <u>Newton's Second Law</u>:

Resultant force (N) ——— $F = ma$ ——— Acceleration (m/s²)

Mass (kg)

EXAMPLE: **A van of mass of 2080 kg has an engine that provides a driving force of 5200 N. At 70 mph the drag force acting on the van is 5148 N. Find its acceleration at 70 mph.**

1) Work out the <u>resultant force</u> on the van. (Drawing a <u>free body diagram</u> may help.)

Resultant force = 5200 − 5148 = 52 N

2) <u>Rearrange</u> $F = ma$ and stick in the <u>values</u> you know.

$a = F \div m$
$= 52 \div 2080 = 0.025$ m/s²

You can use <u>Newton's Second Law</u> to get an idea of the forces involved in everyday transport. Large <u>forces</u> are needed to produce large <u>accelerations</u>:

EXAMPLE: **Estimate the resultant force on a car as it accelerates from rest to a typical speed.**

1) Estimate the <u>acceleration</u> of the car, using <u>typical</u> speeds from page 104.

A typical speed of a car is ~25 m/s. It takes ~10 s to reach this.
So $a = \Delta v \div t = 25 \div 10 = 2.5$ m/s²

The ~ means approximately.

2) <u>Estimate</u> the <u>mass</u> of the car.

Mass of a car is ~1000 kg.

3) Put these numbers into <u>Newton's Second Law</u>.

So using $F = ma = 1000 \times 2.5 = 2500$ N
So the resultant force is ~2500 N

Inertia and Newton's Third Law

Newton's Third Law and inertia sound pretty straightforward, but things can quickly get confusing...

Inertia is the Tendency for Motion to Remain Unchanged

1) Until acted upon by a resultant force, objects at rest stay at rest and objects moving at a steady speed will stay moving at that speed (Newton's First Law). This tendency to continue in the same state of motion is called inertia.

2) An object's inertial mass measures how difficult it is to change the velocity of an object.

3) Inertial mass can be found using Newton's Second Law of $F = ma$ (see the last page). Rearranging this gives $m = F \div a$, so inertial mass is just the ratio of force over acceleration.

Newton's Third Law — Interaction Pairs are Equal and Opposite

Newton's Third Law says:

> When two objects interact, the forces they exert on each other are equal and opposite.

1) If you push something, say a shopping trolley, the trolley will push back against you, just as hard.

2) And as soon as you stop pushing, so does the trolley. Kinda clever really.

3) So far so good. The slightly tricky thing to get your head round is this — if the forces are always equal, how does anything ever go anywhere? The important thing to remember is that the two forces are acting on different objects.

When skater A pushes on skater B, she feels an equal and opposite force from skater B's hand (the 'normal contact' force). Both skaters feel the same sized force, in opposite directions, and so accelerate away from each other.

Skater A will be accelerated more than skater B, though, because she has a smaller mass — remember $a = F \div m$.

An example of Newton's Third Law in an equilibrium situation is a man pushing against a wall. As the man pushes the wall, there is a normal contact force acting back on him. These two forces are the same size. As the man applies a force and pushes the wall, the wall 'pushes back' on him with an equal force.

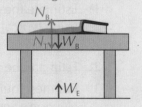

It can be easy to get confused with Newton's Third Law when an object is in equilibrium. E.g. a book resting on a table is in equilibrium. The weight of the book is equal to the normal contact force. The weight of the book pulls it down, and the normal reaction force from the table pushes it up. This is NOT Newton's Third Law. These forces are different types and they're both acting on the book.

The pairs of forces due to Newton's Third Law in this case are:

1) The weight of book is pulled down by gravity from Earth (W_B) and the book also pulls back up on the Earth (W_E).

2) The normal contact force from the table pushing up on the book (N_B) and the normal contact force from the book pushing down on the table (N_T).

Investigating Motion

Here comes another Required Practical. This one's all about testing Newton's Second Law. It uses some nifty bits of kit that you may not have seen before, so make sure you follow the instructions closely.

You can **Investigate** how **Mass** and **Force** Affect Acceleration

It's time for an experiment that tests Newton's 2nd Law, $F = ma$ (p.111).

1) Set up the apparatus shown above. Set up the trolley so it holds a piece of card with a gap in the middle that will interrupt the signal on the light gate twice. If you measure the length of each bit of card that will pass through the light gate and input this into the software, the light gate can measure the velocity for each bit of card. It can use this to work out the acceleration of the trolley.

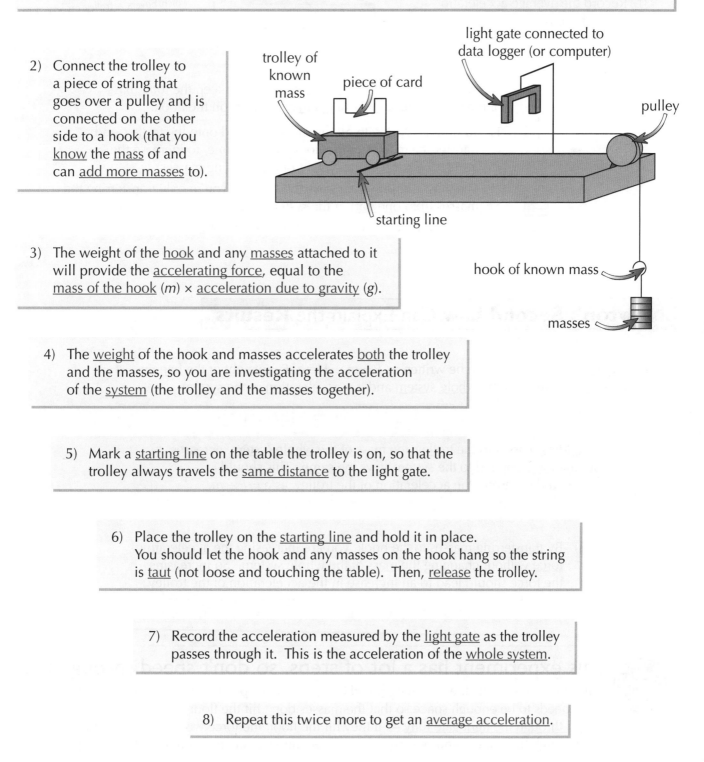

2) Connect the trolley to a piece of string that goes over a pulley and is connected on the other side to a hook (that you know the mass of and can add more masses to).

trolley of known mass

piece of card

light gate connected to data logger (or computer)

pulley

starting line

3) The weight of the hook and any masses attached to it will provide the accelerating force, equal to the mass of the hook (m) × acceleration due to gravity (g).

hook of known mass

masses

4) The weight of the hook and masses accelerates both the trolley and the masses, so you are investigating the acceleration of the system (the trolley and the masses together).

5) Mark a starting line on the table the trolley is on, so that the trolley always travels the same distance to the light gate.

6) Place the trolley on the starting line and hold it in place. You should let the hook and any masses on the hook hang so the string is taut (not loose and touching the table). Then, release the trolley.

7) Record the acceleration measured by the light gate as the trolley passes through it. This is the acceleration of the whole system.

8) Repeat this twice more to get an average acceleration.

Now you've set up the equipment, and you're used to how it works, it's time to start adjusting your variables. Take care with the method here — there are some important points you don't want to miss.

Varying Mass and Force

1) To investigate the effect of mass, add masses to the trolley, one at a time, to increase the mass of the system.
2) Don't add masses to the hook, or you'll change the force.
3) Record the average acceleration for each mass.

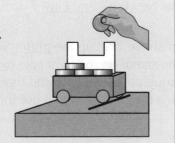

The friction between the trolley and the bench might affect your acceleration measurements. You could use an air track to reduce this friction (a track which hovers a trolley on jets of air).

To investigate the effect of force, you need to keep the total mass of the system the same, but change the mass on the hook.

1) To do this, start with all the masses loaded onto the trolley, and transfer the masses to the hook one at a time, to increase the accelerating force (the weight of the hanging masses).
2) The mass of the system stays the same as you're only transferring the masses from one part of the system (the trolley) to another (the hook).
3) Record the average acceleration for each force.

Newton's Second Law Can Explain the Results

1) Newton's Second Law can be written as $F = ma$. Here, F = weight of the hanging masses, m = mass of the whole system and a = acceleration of the system.

2) By adding masses to the trolley, the mass of the whole system increases, but the force applied to the system stays the same. This should lead to a decrease in the acceleration of the trolley, as $a = F \div m$.

3) By transferring masses to the hook, you are increasing the accelerating force without changing the mass of the whole system. So increasing the force should lead to an increase in the acceleration of the trolley.

This experiment has a lot of steps, so don't speed through it...

Make sure the string is the right length and there's enough space for the hanging masses to fall. There needs to be enough space so that the masses don't hit the floor before the trolley has passed through the light gate fully — if they hit the floor, the force won't be applied the whole way through the trolley's journey, so you won't get an accurate measurement for the speed.

Warm-Up & Exam Questions

Now you've gotten yourself on the right side of the law(s of motion), it's time to put your knowledge on trial. Have a go at cross-examining these questions.

Warm-Up Questions

1) What is the resultant force on an object moving at a constant velocity?

2) Boulders A and B are accelerated from 0 m/s to 5 m/s in 10 s. Boulder A required a force of 70 N, and Boulder B required a force of 95 N. Which boulder has the greater inertial mass?

3) True or False? Two interacting objects exert equal and opposite forces on each other.

4) In a trolley-and-pulley system, as in the practical on page 113, where should you put masses to increase the mass of the system without increasing the force on the trolley?

Exam Questions

1 Dahlia's cricket bat has a mass of 1.2 kg.
She uses it to hit a ball with a mass of 160 g forwards with a force of 500 N. *(Grade 4-6)*

1.1 State the force that the ball exerts on the bat. Explain your answer.

[2 marks]

1.2 Which is greater — the acceleration of the bat or the ball? Explain your answer.

[2 marks]

Figure 1

2 A camper van has a mass of 2500 kg.
It is driven along a straight,
level road at a constant speed *(Grade 6-7)*
of 90.0 kilometres per hour.

2.1 A headwind begins blowing with a force of 200 N, causing the van to slow down.
Calculate the van's deceleration.

[3 marks]

The van begins travelling at a constant speed before colliding with a stationary 10.0 kg traffic cone. The traffic cone accelerates in the direction of the van's motion with an acceleration of 29.0 m/s^2.

2.2 Calculate the force applied to the traffic cone by the van.

[2 marks]

2.3 Calculate the deceleration of the van during the collision.
Assume all of the force applied by the cone to the van causes the deceleration.

[3 marks]

PRACTICAL

3* Stefan is investigating how acceleration varies with force.
He has a 1 kg trolley, attached by a pulley to a 0.5 kg hanging hook. *(Grade 7-9)*
He also has eight 100 g masses. When the hook is released, the trolley rolls
along a table, and passes through a light gate which calculates its acceleration.

Describe an experiment that Stefan can perform using this equipment to investigate the relationship between force and acceleration.

[6 marks]

Stopping Distances

Knowing what affects <u>stopping distances</u> is especially useful for everyday life, as well as the exam.

Stopping Distance is the Sum of Two Distances

1) In an <u>emergency</u> (e.g. a <u>hazard</u> ahead in the road), a driver may perform an <u>emergency stop</u>. This is where <u>maximum force</u> is applied by the <u>brakes</u> in order to stop the car in the <u>shortest possible distance</u>. The <u>longer</u> it takes to perform an <u>emergency stop</u>, the <u>higher the risk</u> of crashing into whatever's in front.

2) The distance it takes to stop a car in an emergency (its <u>stopping distance</u>) is found by:

> Stopping Distance = Thinking Distance + Braking Distance

3) The <u>thinking distance</u> — how far the car travels during the driver's <u>reaction time</u> (the time <u>between</u> the driver <u>seeing</u> a hazard and <u>applying the brakes</u>).

4) The <u>braking distance</u> — the distance taken to stop under the <u>braking force</u> (once the brakes are applied).

Typical car braking distances are: 14 m at 30 mph, 55 m at 60 mph and 75 m at 70 mph.

Many Factors Affect Your Total Stopping Distance

<u>Thinking distance</u> is affected by:
- Your <u>speed</u> — the <u>faster</u> you're going the <u>further</u> you'll travel during the <u>time</u> you take to <u>react</u>.
- Your <u>reaction time</u> — the longer your reaction time (see next page), the longer your <u>thinking distance</u>. This can be affected by <u>tiredness</u>, <u>drugs</u> or <u>alcohol</u>. <u>Distractions</u> can affect your <u>ability</u> to <u>react</u>.

<u>Braking distance</u> is affected by:
- Your <u>speed</u> — for a <u>given</u> braking force, the <u>faster</u> a vehicle travels, the <u>longer</u> it takes to stop (p.118).
- The <u>weather</u> or <u>road surface</u> — if it is <u>wet</u> or <u>icy</u>, or there are <u>leaves</u> or <u>oil</u> on the road, there is <u>less grip</u> (and so less <u>friction</u>) between a vehicle's tyres and the road, which can cause tyres to <u>skid</u>.
- The <u>condition</u> of your <u>tyres</u> — if the tyres of a vehicle are <u>bald</u> (they don't have <u>any tread left</u>) then they cannot <u>get rid of water</u> in wet conditions. This leads to them <u>skidding</u> on top of the water.
- How good your <u>brakes</u> are — if brakes are <u>worn</u> or <u>faulty</u>, they won't be able to apply as much <u>force</u> as well-maintained brakes, which could be dangerous when you need to brake hard.

You need to be able to <u>describe</u> the <u>factors</u> affecting stopping distance and how this affects <u>safety</u> — especially in an <u>emergency</u>.

> For example: <u>Icy</u> conditions increase the chance of <u>skidding</u> (and so increase the stopping distance) so driving <u>too close</u> to other cars in icy conditions is <u>unsafe</u>. The <u>longer</u> your stopping distance, the <u>more space</u> you need to leave <u>in front</u> in order to stop <u>safely</u>.

<u>Speed limits</u> are really important because <u>speed</u> affects the stopping distance so much.

Stopping distance = thinking distance + braking distance.
The exam might ask you to give factors, <u>other than speed</u>, which affect <u>thinking</u> or <u>braking</u> <u>distances</u>, so make sure you know <u>all the factors</u> that affect each of these and <u>what their effects are</u>.

Reaction Times

Reaction times are an <u>important factor</u> in <u>thinking distances</u>. They're also super easy to <u>test</u> for yourself. Read on for a simple <u>experiment</u> you can do in the lab.

You can **Measure** Reaction Times with the **Ruler Drop Test**

<u>Everyone's</u> reaction time is different, but a typical reaction time is between <u>0.2</u> and <u>0.9 s</u> and many different <u>factors</u> affect it (see previous page).

You can do <u>simple experiments</u> to investigate your reaction time, but as reaction times are <u>so short</u>, you haven't got a chance of measuring one with a <u>stopwatch</u>. One way of measuring reaction times is to use a <u>computer-based test</u> (e.g. <u>clicking a mouse</u> when the screen changes colour). Another is the <u>ruler drop test</u>. Here's how to carry it out:

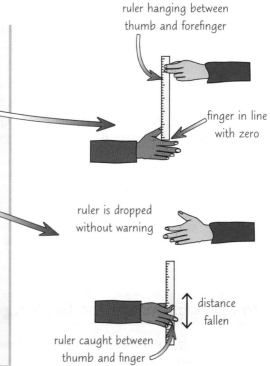

ruler hanging between thumb and forefinger

finger in line with zero

ruler is dropped without warning

distance fallen

ruler caught between thumb and finger

1) Sit with your arm resting on the edge of a table (this should stop you moving your arm up or down during the test). Get someone else to hold a ruler so it <u>hangs between</u> your thumb and forefinger, lined up with <u>zero</u>. You may need a <u>third person</u> to be at <u>eye level with the ruler</u> to check it's lined up.

2) Without giving any warning, the person holding the ruler should <u>drop it</u>. Close your thumb and finger to try to <u>catch the ruler as quickly as possible</u>.

3) The measurement on the ruler at the point where it is caught is <u>how far</u> the ruler dropped in the time it takes you to react.

4) The <u>longer</u> the <u>distance</u>, the <u>longer</u> the <u>reaction time</u>.

5) You can calculate <u>how long</u> the ruler falls for (the <u>reaction</u> time) because <u>acceleration due to gravity is constant</u> (roughly 9.8 m/s²).

E.g. say you catch the ruler at 20 cm. From p.105 you know: $\underline{v^2 - u^2 = 2as}$.

$u = 0$, $a = 9.8$ m/s² and $s = 0.2$ m, so: $v = \sqrt{2 \times 9.8 \times 0.2 + 0}$ = <u>2.0 m/s (to 2 s.f.)</u>

v is equal to the <u>change in velocity</u> of the ruler.

From page 105 you also know: $\underline{a = \Delta v \div t}$ so $\underline{t = \Delta v \div a = 2.0 \div 9.8 = 0.2\ s}$ (to 1 s.f.). This gives your <u>reaction time</u>.

---

6) It's <u>pretty hard</u> to do this experiment <u>accurately</u>, so you should do a lot of <u>repeats</u>. The results will be better if the ruler falls <u>straight down</u> — you might want to add a <u>blob of modelling clay</u> to the bottom to stop it from waving about.

7) Make sure it's a <u>fair test</u> — use the <u>same ruler</u> for each repeat, and have the <u>same person</u> dropping it.

8) You could try to investigate some factors affecting reaction time, e.g. you could introduce <u>distractions</u> by having some <u>music</u> playing or by having someone <u>talk to you</u> while the test takes place (see the previous page for more on the factors affecting reaction time).

9) Remember to still do lots of <u>repeats</u> and calculate the <u>mean</u> reaction time with distractions, which you can <u>compare</u> to the mean reaction time <u>without</u> distractions.

Braking Distances

So you know the basics of stopping distances now, but how do the brakes actually work to slow down a car? Well, it's all down to friction and transferring energy away from the wheels to the brakes.

Braking Relies on Friction Between the Brakes and Wheels

1) When the brake pedal is pushed, this causes brake pads to be pressed onto the wheels. This contact causes friction, which causes work to be done.

2) The work done between the brakes and the wheels transfers energy from the kinetic energy stores of the wheels to the thermal energy stores of the brakes. The brakes increase in temperature.

3) The faster a vehicle is going, the more energy it has in its kinetic store, so the more work needs to be done to stop it. This means that a greater braking force is needed to make it stop within a certain distance.

4) A larger braking force means a larger deceleration. Very large decelerations can be dangerous because they may cause brakes to overheat (so they don't work as well) or could cause the vehicle to skid.

You Can Estimate the Braking Force Required to Stop

You can estimate the braking force required to make a vehicle decelerate and come to a stop. As you're only estimating the force, this is a place where you may well need to use typical values:

EXAMPLE:

A car travelling at a typical speed makes an emergency stop to avoid hitting a hazard 25 m ahead.
Estimate the braking force needed to produce this deceleration.

For a refresher on typical speed values, head back to page 104.

1) Assume the deceleration is uniform, and rearrange $v^2 - u^2 = 2as$ to find the deceleration.

$v = \sim25$ m/s $m = \sim1000$ kg.
$a = (v^2 - u^2) \div 2s = (0^2 - 25^2) \div (2 \times 25) = -12.5$

2) Then use $F = ma$, with $m = \sim1000$ kg.

$F = ma$
$F = 1000 \times 12.5 = 12\ 500$ N, so F is $\sim12\ 500$ N

Typical values, always there when you need them...

REVISION TIP

Make sure you memorise the typical speed values on page 104. It shouldn't matter if they're slightly off, but they need to be of roughly the right size, or your calculations won't make sense. If you're asked about a vehicle that you don't know the typical speed or mass of, try to use those you do know as a guide. For example, a bus will have a greater typical mass than a car (1000 kg) but a slightly smaller typical speed than a car (25 m/s), as it's bigger and travels on slower routes.

Speed and Stopping Distances

As you've seen, <u>stopping distance</u> is made up of <u>two</u> components, <u>thinking distance</u> and <u>braking distance</u>. And there's one <u>factor</u> which affects <u>both</u> components significantly — your <u>speed</u>.

Leave Enough **Space** to **Stop**

1) The figures on the right for <u>typical stopping distances</u> are from the <u>Highway Code</u>.

2) To <u>avoid an accident</u>, drivers need to leave <u>enough space</u> between their car and the one in front so that if they had to <u>stop suddenly</u> they would have time to do so <u>safely</u>. 'Enough space' means the <u>stopping distance</u> for whatever speed they're going at.

3) So even at <u>30 mph</u>, you should drive no closer than <u>6 or 7 car lengths</u> away from the car in front.

4) <u>Speed limits</u> are really important because <u>speed</u> affects the stopping distance so much.

Don't forget — things like bad weather and road conditions will make stopping distances even longer (see page 116).

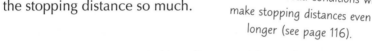

Speed Affects Braking Distance
More Than Thinking Distance

1) As a car <u>speeds up</u>, the <u>thinking distance increases</u> at the <u>same rate</u> as speed. The graph is <u>linear</u> (a straight line).

2) This is because the thinking time (how long it takes the driver to apply the brakes) stays pretty <u>constant</u> — but the higher the speed, the more distance you cover in that time.

3) <u>Braking distance</u>, however, increases <u>faster</u> the more you speed up. The <u>work done</u> to stop the car is <u>equal</u> to the energy in the car's <u>kinetic energy store</u> ($\frac{1}{2}mv^2$). So as speed doubles, the kinetic energy increases <u>4-fold</u> (2^2), and so the <u>work done</u> to stop the car increases 4-fold. Since <u>$W = Fs$</u> (p.90) and the <u>braking force</u> is <u>constant</u>, the <u>braking distance increases 4-fold</u>.

4) <u>Stopping distance</u> is a <u>combination</u> of these two distances (p.116) so the graph of <u>speed</u> against <u>stopping distance</u> for a car looks like this:

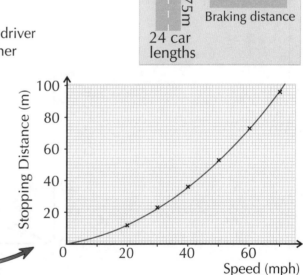

5) You need to be able to <u>interpret</u> graphs like this for a <u>range</u> of vehicles — they're all a similar shape.

EXAMPLE: **Below is a graph of stopping distance against speed for two vehicles, A and B. Compare the stopping distance for both vehicles at a speed of 40 mph.**

1) Read off the graph to find the <u>stopping distance</u> for each vehicle at <u>40 mph</u>.

Vehicle A stopping distance = 34 m
Vehicle B stopping distance = 41 m

2) Find the <u>difference</u> between these two values.

41 − 34 = 7 m

So the stopping distance for vehicle B is 7 m longer than for vehicle A.

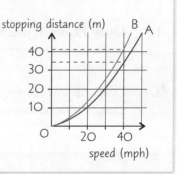

Warm-Up & Exam Questions

Time to apply the brakes for a second and put your brain through an MOT. Try out these questions. If you can handle these, your exam should be clear of hazards.

Warm-Up Questions

1) What is meant by 'thinking distance'?
2) What must be added to the thinking distance to find the total stopping distance of a car?
3) Give an example of how poor weather can affect your ability to stop a car before hitting a hazard.
4) Describe an experiment you could carry out, using a ruler, to measure the reaction time of an individual.
5) What energy transfer occurs when a car brakes?

Exam Question

1   **Figure 1** shows how thinking distance and stopping distance vary with speed for a car travelling on a clear day on a dry road.

Grade 4-6

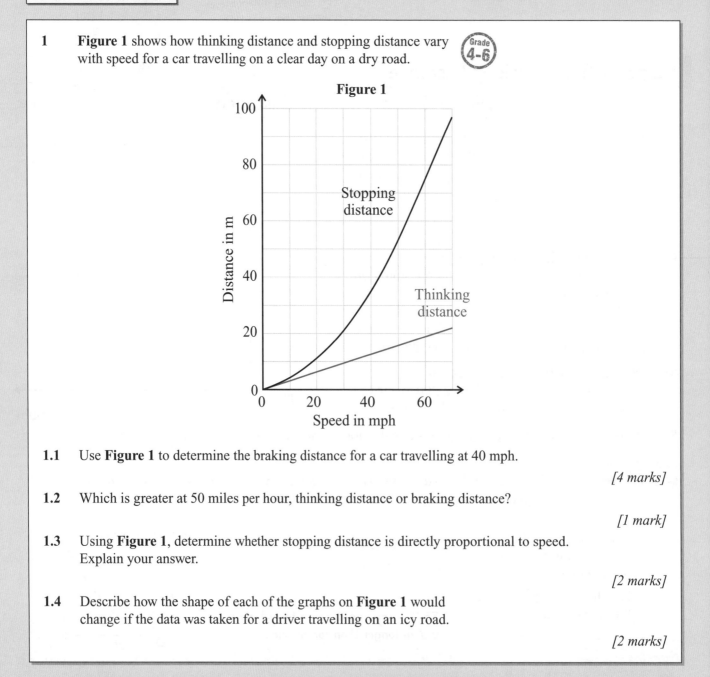

Figure 1

1.1   Use **Figure 1** to determine the braking distance for a car travelling at 40 mph.

[4 marks]

1.2   Which is greater at 50 miles per hour, thinking distance or braking distance?

[1 mark]

1.3   Using **Figure 1**, determine whether stopping distance is directly proportional to speed. Explain your answer.

[2 marks]

1.4   Describe how the shape of each of the graphs on **Figure 1** would change if the data was taken for a driver travelling on an icy road.

[2 marks]

Momentum

A large rugby player running very fast has much more momentum than a skinny bloke out for a Sunday afternoon stroll. Momentum's something that all moving objects have, so you better get your head around it.

Momentum = Mass × Velocity

Momentum is mainly about how much 'oomph' an object has. It's a property that all moving objects have.

1) The greater the mass of an object, or the greater its velocity, the more momentum the object has.

2) Momentum is a vector quantity — it has size and direction.

3) You can work out the momentum of an object using:

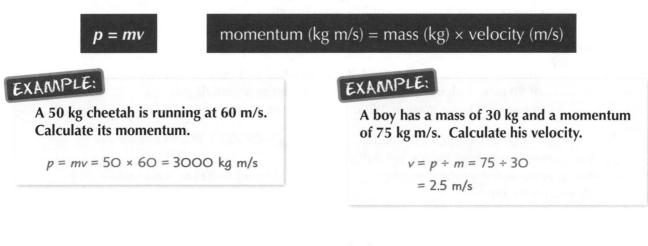

$$p = mv$$

momentum (kg m/s) = mass (kg) × velocity (m/s)

EXAMPLE:

A 50 kg cheetah is running at 60 m/s. Calculate its momentum.

$p = mv = 50 \times 60 = 3000$ kg m/s

EXAMPLE:

A boy has a mass of 30 kg and a momentum of 75 kg m/s. Calculate his velocity.

$v = p \div m = 75 \div 30$

$= 2.5$ m/s

Momentum **Before** = Momentum **After**

In a closed system, the total momentum before an event (e.g. a collision) is the same as after the event. This is called conservation of momentum.

A closed system is just a fancy way of saying that no external forces act.

In snooker, balls of the same size and mass collide with each other. Each collision is an event where the momentum of each ball changes, but the overall momentum stays the same (momentum is conserved).

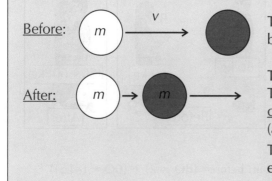

Before:

After:

The red ball is stationary, so it has zero momentum. The white ball is moving with a velocity v, so has a momentum of $p = mv$.

The white ball hits the red ball, causing it to move. The red ball now has momentum. The white ball continues moving, but at a much smaller velocity (and so a much smaller momentum).

The combined momentum of the red and white ball is equal to the original momentum of the white ball, mv.

A moving car hits into the back of a parked car. The crash causes the two cars to lock together, and they continue moving in the direction that the original moving car was travelling, but at a lower velocity.

Before: The momentum was equal to mass of moving car × its velocity.
After: The mass of the moving object has increased, but its momentum is equal to the momentum before the collision. So an increase in mass causes a decrease in velocity.

If the momentum before an event is zero, then the momentum after will also be zero.
E.g. in an explosion, the momentum before is zero. After the explosion, the pieces fly off in different directions, so that the total momentum cancels out to zero.

Momentum

You can use the <u>equation</u> for <u>momentum</u>, along with the <u>conservation of momentum principle</u>, to <u>calculate</u> changes in <u>mass</u> and <u>velocity</u> in interactions. It's all about '<u>momentum before = momentum after</u>'.

You Can Use **Conservation of Momentum**

to **Calculate Velocities** or **Masses**

You've already seen that <u>momentum is conserved</u> in a <u>closed system</u> (see last page).
You can use this to help you calculate things
like the <u>velocity</u> or <u>mass</u> of objects in an event (e.g. a collision).

EXAMPLE:

Misha fires a paintball gun. A 3.0 g paintball is fired at a velocity of 90 m/s. Calculate the velocity at which the paintball gun recoils if it has a mass of 1.5 kg. Momentum is conserved.

The word recoil means to move backwards.

1) Calculate the <u>momentum</u> of the <u>pellet</u>.

$p = 0.003 \times 90 = 0.27$ kg m/s

2) The momentum before the gun is fired is <u>zero</u>.
   This is equal to the <u>total</u> momentum after the collision.

Momentum before = momentum after

$0 = 0.27 + (1.5 \times v)$

3) The momentum of the <u>gun</u> is 1.5 × v.

4) <u>Rearrange</u> the equation to find the <u>velocity</u> of the gun. The <u>minus sign</u> shows the gun is travelling in the <u>opposite direction</u> to the bullet.

$v = -(0.27 \div 1.5)$
$= -0.18$ m/s

EXAMPLE:

Two skaters, Skater A and Skater B, approach each other, collide and move off together as shown in the image on the right. At what velocity do they move after the collision?

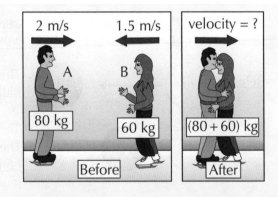

2 m/s   1.5 m/s   velocity = ?

A   B

80 kg   60 kg   (80 + 60) kg

Before   After

1) Choose which direction is <u>positive</u>.
   I'll say "<u>positive</u>" means "<u>to the right</u>".

2) <u>Total momentum before</u> collision
   = momentum of A + momentum of B

Momentum before= (80 × 2) + (60 × (−1.5))
                    = 70 kg m/s

3) <u>Total momentum after</u> collision
   = momentum of A and B together

Momentum after = 140 × v

4) Set momentum before equal to momentum after, and rearrange for the answer.

$140v = 70$, so $v = 70 \div 140$
$v = 0.5$ m/s to the right

EXAM TIP

Momentum questions may need you to analyse a scenario...

Make sure you <u>read</u> any momentum questions <u>carefully</u>. You need to identify what the <u>objects</u> and <u>momentum</u> were <u>before the interaction</u>, and what they are <u>after the interaction</u>. The question may not be a scenario you're familiar with, so you'll need to work out what's going on.

Changes in Momentum

When a force acts on an object, it causes the object to change momentum.
The bigger the force, the faster the change in momentum. And the reverse is true too.

Forces Cause a Change in Momentum

1) You know that when a non-zero resultant force acts on a moving object (or an object that can move), it causes its velocity to change (p.111). This means that there is a change in momentum.

2) You also know that $F = ma$ and that a = change in velocity ÷ change in time.

3) So $F = m \times \frac{v-u}{t}$, which can also be written as:

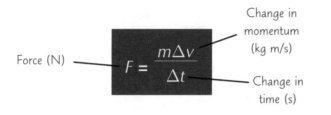

Force (N) — $F = \dfrac{m\Delta v}{\Delta t}$ — Change in momentum (kg m/s)

Change in time (s)

Equations tell you how variables are related. You should be able to use them to work out how changing one will affect the other.

4) The force causing the change is equal to the rate of change of momentum.

5) A larger force means a faster change of momentum.

6) Likewise, if someone's momentum changes very quickly (like in a car crash), the forces on the body will be very large and more likely to cause injury.

7) This is why cars are designed to slow people down over a longer time when they have a crash — the longer it takes for a change in momentum, the smaller the rate of change of momentum, and so the smaller the force. Smaller forces mean the injuries are likely to be less severe.

Cars have many safety features, such as:
Crumple zones crumple on impact, increasing the time taken for the car to stop.
Seat belts stretch slightly, increasing the time taken for the wearer to stop.
Air bags inflate before you hit the dashboard of a car. The compressing air inside it slows you down more gradually than if you had just hit the hard dashboard.

Helmets, e.g. bike helmets contain a crushable layer of foam which helps to lengthen the time taken for your head to stop in a crash. This reduces the impact on your brain.

Crash mats and cushioned playground flooring increase the time taken for you to stop if you fall on them. This is because they are made from soft, compressible (squishable) materials.

Many safety features decrease rate of change of momentum...

Knowledge of the connection between force and the rate of change of momentum has allowed us to develop a range of safety features designed to minimise injury in crashes and collisions. This is a clear example of science being applied to develop useful new technologies and devices.

Warm-Up & Exam Questions

Here it is, the final hurdle before the end of the topic! Keep up your momentum, and throw yourself into these questions. You'll be over the finish line before you know it.

Warm-Up Questions

1) Calculate the momentum of a 2.5 kg rabbit running through a garden at 10 m/s.
2) What is meant by the conservation of momentum?
3) What is the total momentum before and after an explosion?
4) How does increasing the time over which an object changes momentum change the force on it?

Exam Questions

1   An 60 kg gymnast lands on a crash mat. When they hit the crash mat, they are moving at 5.0 m/s and come to a stop in a period of 1.2 seconds (after which their momentum is zero).   *Grade 6-7*

1.1   State the equation linking momentum, mass and velocity.

[1 mark]

1.2   Calculate the momentum of the gymnast immediately before they hit the crash mat.

[2 marks]

1.3   Calculate the size of the average force acting on the gymnast as they land on the crash mat. Use the correct equation from the Physics Equation Sheet on the inside back cover.

[2 marks]

2   In a demolition derby, cars drive around an arena and crash into each other.   *Grade 6-7*

2.1   One car has a mass of 650 kg and a velocity of 15.0 m/s. Calculate the momentum of the car.

[2 marks]

2.2   The car collides head-first with another car with a mass of 750 kg. The two cars stick together. Calculate the combined velocity of the two cars immediately after the collision if the other car had a velocity of 10.0 m/s before the collision. Give your answer to 3 significant figures.

[5 marks]

2.3   The cars have crumple zones at the front of the car that crumple on impact. Explain how a crumple zone reduces the forces acting on a driver during a collision.

[2 marks]

3   A fast-moving neutron collides with a uranium-235 atom and bounces off. **Figure 1** shows the particles before and after the collision.   *Grade 7-9*

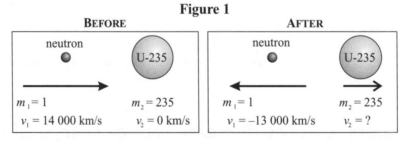

Figure 1

Calculate the velocity of the U-235 atom after the collision. Give your answer to 3 significant figures

[5 marks]

Revision Summary for Topic 5

That wraps up <u>Topic 5</u> — time to put yourself to the test and find out <u>how much you really know</u>.
* Try these questions and <u>tick off each one</u> when you <u>get it right</u>.
* When you've done <u>all the questions</u> under a heading and are <u>completely happy</u> with it, tick it off.

Forces and Work Done (p.87-91) ☑

1) Explain the difference between scalar and vector quantities, and contact and non-contact forces. ☑
2) What is the formula for calculating the weight of an object? ☑
3) What is a free body diagram? ☑
4) Give the formula for calculating the work done by a force. ☑
5) Describe the forces acting on an object in equilibrium. ☑

Stretching and Turning (p.93-98) ☑

6) What is the difference between an elastic and an inelastic deformation? ☑
7) How do you find the spring constant from a linear force-extension graph? ☑
8) What is the area under the linear part of a force-extension graph of an object equal to? ☑
9) How does the moment of a force vary with the distance from the pivot at which the force is applied? ☑
10) If a seesaw is balanced, what can you say about the moments? ☑
11) What happens to the direction of rotation each time a force is transmitted between gears? ☑

Pressure (p.100-102) ☑

12) State an equation for calculating the pressure at the surface of a fluid. Give the units of pressure. ☑
13) Explain why the pressure increases as you go deeper into a column of a liquid. ☑
14) What causes an object to float? ☑
15) Explain how and why atmospheric pressure varies with height. ☑

Motion (p.104-114) ☑

16) What is the difference between displacement and distance? ☑
17) Define acceleration in terms of velocity and time. ☑
18) What does the term 'uniform acceleration' mean? ☑
19) What does the gradient represent for: a) a distance-time graph? b) a velocity-time graph? ☑
20) What is terminal velocity? Why do objects reach terminal velocity? ☑
21) State Newton's three laws of motion. ☑
22) What is inertia? ☑

Car Safety and Momentum (p.116-123) ☑

23) What is the stopping distance of a vehicle? How can it be calculated? ☑
24) Give two things that affect a person's reaction time. ☑
25) What is an average reaction time? ☑
26) Give two examples of methods that could be used to test a person's reaction time. ☑
27) State four things that can affect the braking distance of a vehicle. ☑
28) Is momentum a vector or a scalar quantity? ☑
29) Explain how car safety features use momentum and forces to reduce the risk of injury to passengers. ☑

Wave Basics

Waves <u>transfer energy</u> from one place to another <u>without</u> transferring any <u>matter</u> (stuff).

Waves Transfer **Energy** in the **Direction** they are **Travelling**

When waves travel through a medium, the <u>particles</u> of the medium <u>oscillate</u> and <u>transfer energy</u> between each other. BUT overall, the particles stay in the <u>same place</u> — <u>only energy</u> is transferred.

> For example, if you drop a twig into a calm pool of water, <u>ripples</u> form on the water's surface. The ripples <u>don't</u> carry the <u>water</u> (or the twig) away with them though.
>
> Similarly, if you strum a <u>guitar string</u> and create <u>sound waves</u>, the sound waves don't carry the <u>air</u> away from the guitar and create a <u>vacuum</u>.

Waves have **Amplitude**, **Wavelength** and **Frequency**

1) The <u>amplitude</u> of a wave is the <u>maximum displacement</u> of a point on the wave from its <u>undisturbed position</u>.

2) The <u>wavelength</u> is the distance between the <u>same point</u> on two <u>adjacent</u> waves (e.g. between the <u>trough</u> of one wave and the <u>trough</u> of the wave <u>next to it</u>).

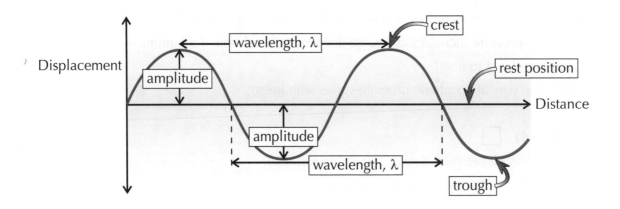

3) <u>Frequency</u> is the <u>number of complete waves</u> passing a certain point <u>per second</u>. Frequency is measured in <u>hertz</u> (Hz). 1 Hz is <u>1 wave per second</u>.

The period of a wave is the amount of <u>time</u> it takes for a <u>full cycle</u> of the wave to pass a point. You can find it from the <u>frequency</u> of the wave using the <u>formula</u>:

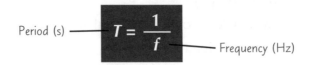

Period (s) —— $T = \dfrac{1}{f}$ —— Frequency (Hz)

The only thing a wave transfers is energy...

It's <u>really</u> important that you understand this stuff <u>really</u> well, or the rest of this topic will simply be a blur. Make sure you can sketch the <u>wave diagram</u> above and can <u>label</u> all the features from memory. Then check you know all the <u>definitions</u> and the <u>equation</u> linking period and frequency.

Transverse and Longitudinal Waves

All waves are either transverse or longitudinal. Read on to find out more...

Transverse Waves Have Perpendicular Vibrations

In transverse waves, the oscillations (vibrations) are perpendicular (at 90°) to the direction of energy transfer.
A spring wiggled from side to side gives a transverse wave:

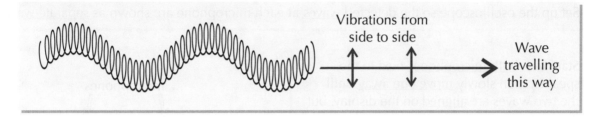

Most waves are transverse, including:

1) All electromagnetic waves, e.g. light (page 136).
2) Ripples and waves in water (page 128).
3) A wave on a string (page 129).

Water waves, shock waves and waves in springs and ropes are all examples of mechanical waves.

Longitudinal Waves Have Parallel Vibrations

In longitudinal waves, the oscillations are parallel to the direction of energy transfer.
If you push the end of a spring, you get a longitudinal wave.

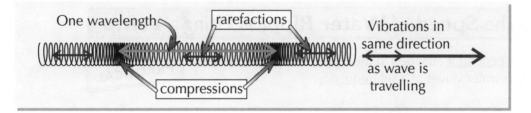

Other examples of longitudinal waves are:

1) Sound waves in air, ultrasound.
2) Shock waves, e.g. some seismic waves (see page 155).

Wave Speed = Frequency × Wavelength

The wave speed is the speed at which energy is being transferred (or the speed the wave is moving at).
The wave equation applies to all waves:

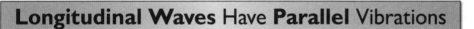

Wave speed (m/s) —— $v = f\lambda$ —— Wavelength (m)
Frequency (Hz)

EXAMPLE:

**A radio wave has a frequency of 12.0×10^6 Hz. Find its wavelength.
(The speed of radio waves in air is 3.0×10^8 m/s.)**

1) To find λ, you need to rearrange the equation $v = f\lambda$. $\lambda = v \div f$
2) Substitute in the values for v and f to calculate λ. $= (3.0 \times 10^8) \div (12.0 \times 10^6) = 25$ m

Experiments with Waves

Measuring the <u>speed of waves</u> isn't that simple. It calls for crafty methods...

Use an **Oscilloscope** to Measure the **Speed** of **Sound**

By attaching a <u>signal generator</u> to a speaker you can generate sounds with a specific <u>frequency</u>.
You can use <u>two microphones</u> and an <u>oscilloscope</u> to find the <u>wavelength</u> of the sound waves generated.

1) Set up the oscilloscope so the <u>detected waves</u> at each microphone are shown as <u>separate waves</u>.

2) Start with <u>both microphones</u> next to the speaker, then slowly <u>move one away</u> until the two waves are <u>aligned</u> on the display, but have moved <u>exactly one wavelength apart</u>.

3) Measure the <u>distance between the microphones</u> to find one <u>wavelength</u> (λ).

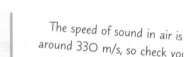

microphones   oscilloscope

wavelength   waves line up

speaker attached
to signal generator

4) You can then use the formula <u>$v = f\lambda$</u> (see last page) to find the <u>speed</u> (v) of the <u>sound waves</u> passing through the <u>air</u> — the <u>frequency</u> (f) is whatever you set the <u>signal generator</u> to (around 1 kHz is sensible).

The speed of sound in air is around 330 m/s, so check your results roughly agree with this.

Measure the **Speed** of **Water Ripples** Using a **Lamp**

Using a <u>signal generator</u> attached to the <u>dipper</u> of a <u>ripple tank</u>, you can create water waves at a <u>set frequency</u>.

PRACTICAL

1) Dim the lights and <u>turn on</u> the <u>lamp</u> — you'll see a <u>wave pattern</u> made by the shadows of the <u>wave crests</u> on the screen below the tank.

2) The distance between each shadow line is equal to one wavelength (p.126). Measure the <u>distance</u> between shadow lines that are 10 wavelengths apart, then <u>divide</u> this distance by 10 to find the <u>average wavelength</u>. This is a <u>suitable method</u> for measuring <u>small</u> wavelengths.

3) If you're struggling to measure the distance, you could take a <u>photo</u> of the <u>shadows and ruler</u>, and find the wavelength from the photo instead.

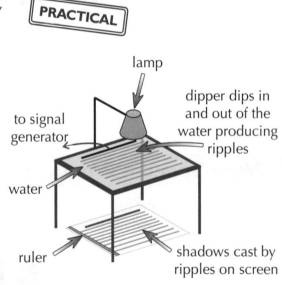

lamp

to signal
generator

dipper dips in
and out of the
water producing
ripples

water

ruler

shadows cast by
ripples on screen

4) Use $v = f\lambda$ to calculate the <u>speed</u> of the waves.

5) This set-up is <u>suitable</u> for investigating waves, because it allows you to measure the wavelength without <u>disturbing</u> the waves.

Experiments with Waves

One more <u>wave experiment</u> coming up. This time, it's to do with <u>waves on strings</u>.

You can Use the **Wave Equation** for Waves on **Strings**

In this practical, you create a wave on a string. Again, you use a <u>signal generator</u>, but this time you attach it to a <u>vibration transducer</u> which converts the signals to vibrations.

1) Set up the equipment shown below, then <u>turn on</u> the signal generator and vibration transducer. The string will start to <u>vibrate</u>.

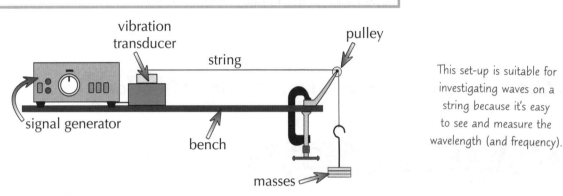

This set-up is suitable for investigating waves on a string because it's easy to see and measure the wavelength (and frequency).

2) You can adjust the <u>frequency</u> setting on the signal generator to change the <u>length</u> of the wave created on the string. You should keep adjusting the frequency of the signal generator until there appears to be a <u>clear wave</u> on the string. This happens when a <u>whole number</u> of half-wavelengths fit exactly on the string (you want at least four or five half-wavelengths ideally). The frequency you need will depend on the <u>length</u> of string between the <u>pulley</u> and the <u>transducer</u>, and the <u>masses</u> you've used.

3) You need to measure the <u>wavelength</u> of the wave. The best way to do this <u>accurately</u> is to measure the length of all the <u>half-wavelengths</u> on the string <u>in one go</u>, then <u>divide</u> by the total number of half-wavelengths to get the <u>mean half-wavelength</u> (see p.8 for more on calculating the mean). You can then <u>double</u> this value to get a <u>full wavelength</u>.

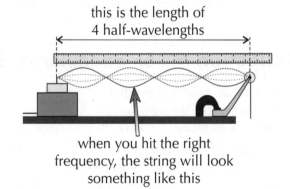

this is the length of 4 half-wavelengths

when you hit the right frequency, the string will look something like this

4) The <u>frequency</u> of the wave is whatever the <u>signal generator</u> is set to.

5) You can find the <u>speed</u> of the wave using $v = f\lambda$.

Learn the methods for all these practicals...

These experiments seem complicated, but they all have a few things <u>in common</u>. First, you set the <u>frequency</u> on the signal generator, then find the length of the resulting wave (this tends to be the fiddly bit). You can then use the <u>equation</u> $v = f\lambda$ to find the <u>wave speed</u>. That's about it.

Reflection

You should be pretty familiar with the idea of <u>reflection</u>. Of course, there's a bit more too it than you'd think.

All Waves Can be **Absorbed, Transmitted** or **Reflected**

When waves arrive at a <u>boundary</u> between two <u>different materials</u>, <u>three</u> things can happen:

1) The waves are <u>absorbed</u> by the material the wave is trying to cross into — this <u>transfers energy</u> to the <u>material's energy stores</u> (this is how a microwave works, see page 138).

2) The waves are <u>transmitted</u> — the waves <u>carry on travelling</u> through the new material. This often leads to <u>refraction</u> (see the next page).

3) The waves are <u>reflected</u> — there's more on this below.

What actually happens depends on the <u>wavelength</u> of the wave and the <u>properties</u> of the <u>materials</u> involved.

You Can Draw a Simple **Ray Diagram** for **Reflection**

1) There's <u>one simple rule</u> to learn for <u>all</u> reflected waves:

> Angle of incidence = Angle of reflection

A ray is a line showing the path a wave travels in. It's perpendicular to a wave's wave fronts (see p.131). Rays are always drawn as straight lines.

2) <u>The angle of incidence</u> is the angle between the <u>incoming wave</u> and the <u>normal</u>.

3) <u>The angle of reflection</u> is the angle between the <u>reflected wave</u> and the normal.

4) The <u>normal</u> is an <u>imaginary line</u> that's <u>perpendicular</u> (at right angles) to the <u>surface</u> at the <u>point of incidence</u> (the point where the wave <u>hits</u> the boundary).

5) The normal is usually shown as a <u>dotted line</u>.

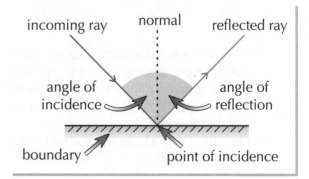

Reflection can be **Specular** or **Diffuse**

1) Waves are reflected at <u>different boundaries</u> in <u>different ways</u>. (There's an investigation on this on page 133.)

2) <u>Specular reflection</u> happens when a wave is reflected in a <u>single direction</u> by a <u>smooth</u> surface. E.g. when <u>light</u> is reflected by a <u>mirror</u> you get a nice <u>clear reflection</u>.

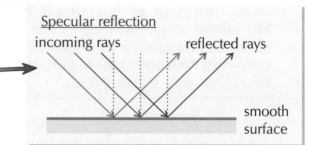

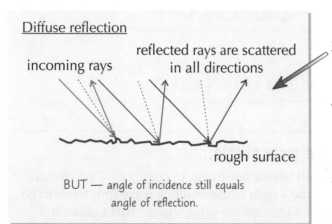

BUT — angle of incidence still equals angle of reflection.

3) <u>Diffuse reflection</u> is when a wave is reflected by a <u>rough</u> surface (e.g. a <u>piece of paper</u>) and the reflected rays are <u>scattered</u> in <u>lots of different directions</u>.

4) This happens because the <u>normal</u> is <u>different</u> for each incoming ray, which means that the <u>angle of incidence is different</u> for each ray. The rule of angle of incidence = angle of reflection <u>still applies</u>.

5) When <u>light</u> is reflected by a rough surface, the surface appears <u>matte</u> (<u>not shiny</u>) and you <u>don't</u> get a <u>clear reflection</u> of objects.

Refraction

Refraction is when light waves are <u>bent</u> when they enter a <u>new media</u> (which is a posh word for material).

Refraction — Waves **Changing Direction** at a **Boundary**

1) When a wave crosses a <u>boundary</u> between materials at an <u>angle</u>, it <u>changes direction</u> — it's <u>refracted</u>.

2) <u>How much</u> it's refracted by depends on how much the wave <u>speeds up</u> or <u>slows down</u>, which usually depends on the <u>density</u> of the two materials (usually the <u>higher</u> the density of a material, the <u>slower</u> a wave travels through it). *See page 136 for more on different waves.*

3) If a wave crosses a boundary and <u>slows down</u> it will bend <u>towards</u> the <u>normal</u>. If it crosses into a material and <u>speeds up</u> it will bend <u>away</u> from the normal.

4) The <u>wavelength</u> of a wave changes when it is refracted, but the <u>frequency stays the same</u>.

5) If the wave is travelling <u>along the normal</u> it will <u>change speed</u>, but it's <u>NOT refracted</u>.

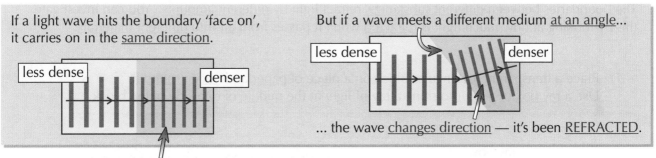

If a light wave hits the boundary 'face on', it carries on in the <u>same direction</u>.

less dense | denser

But if a wave meets a different medium <u>at an angle</u>...

less dense | denser

... the wave <u>changes direction</u> — it's been <u>REFRACTED</u>.

The wave fronts being closer together shows a change in wavelength (and so a change in velocity).

6) The <u>optical density</u> of a material is a measure of <u>how quickly light</u> can travel through it — the <u>higher</u> the optical density, the <u>slower</u> light waves travel through it.

You can **Construct** a **Ray Diagram** to show **Refraction**

1) First, draw the <u>boundary</u> between your two materials and the <u>normal</u> (a line at 90° to the boundary).

incoming ray

angle of incidence

boundary

normal

2) Draw an incident <u>ray</u> that <u>meets</u> the <u>normal</u> at the <u>boundary</u>. The angle <u>between</u> the <u>ray</u> and the <u>normal</u> is the <u>angle of incidence</u>. (If you're <u>given</u> this angle, make sure to draw it <u>carefully</u> with a <u>protractor</u>.)

3) Now draw the <u>refracted ray</u> on the other side of the boundary. If the second material is <u>optically denser</u> than the first, the refracted ray <u>bends towards</u> the normal (like on the right). The <u>angle</u> between the <u>refracted</u> ray and the <u>normal</u> (the angle of <u>refraction</u>) is <u>smaller</u> than the <u>angle of incidence</u>. If the second material is <u>less optically dense</u>, the angle of refraction is <u>larger</u> than the angle of incidence.

angle of refraction

refracted ray

REVISION TIP

Hitting a boundary at an angle leads to refraction...

If you can't remember <u>which way</u> a wave bends when it hits an optically denser material at an angle, imagine a skier skiing from some nice smooth snow onto some rough ground at an angle. The ski hitting the rough ground first <u>slows down first</u>, so they will swing towards that side.

PRACTICAL | Investigating Light

This is the first of two investigations in this topic that you'll come across. They're both to do with the <u>behaviour of light</u>. For both of them, you'll need a <u>ruler</u>, <u>protractor</u> and a <u>nice sharp pencil</u>.

You Need to Do **Light Experiments** in a **Dim Room**

1) Both this experiment and the one on the next page use <u>rays of light</u>, so it's best to do these experiments in a <u>dim room</u> so you can <u>clearly</u> see paths of the rays of light.

2) They also both use either a <u>ray box</u> or a <u>laser</u> to produce <u>thin</u> rays of light. This is so you can <u>trace</u> the paths of the rays more <u>accurately</u>, meaning more exact <u>angle measurements</u>.

You Can Use **Transparent Materials** to Investigate **Refraction**

The boundaries between <u>different substances</u> refract light by <u>different amounts</u>. You can investigate this by looking at how much light is <u>refracted</u> when it passes from <u>air</u> into <u>different materials</u>.

1) Place a transparent rectangular block on a piece of <u>paper</u> and <u>trace around it</u>. Use a <u>ray box</u> or a <u>laser</u> to shine a ray of light at the <u>middle</u> of one side of the block.

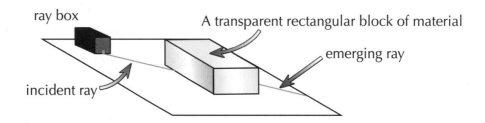

ray box

A transparent rectangular block of material

emerging ray

incident ray

2) <u>Trace</u> the <u>incident ray</u> and <u>mark</u> where the light ray <u>emerges</u> on the other side of the block. Remove the block and, with a <u>straight line</u>, <u>join up</u> the <u>incident ray</u> and the emerging point to show the path of the <u>refracted ray</u> through the block.

3) Draw the <u>normal</u> at the <u>point</u> where the light ray entered the block. Use a protractor to measure the <u>angle</u> between the <u>incident</u> ray and the <u>normal</u> (the <u>angle of incidence</u>, *I*) and the angle between the <u>refracted</u> ray and the <u>normal</u> (the <u>angle of refraction</u>, *R*).

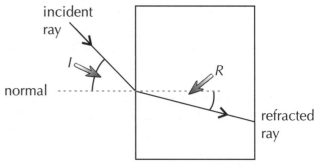

incident ray

normal

I

R

refracted ray

4) <u>Repeat</u> this experiment using rectangular blocks made from different materials, keeping the <u>incident angle</u> the <u>same</u> throughout.

You should find that the angle of refraction <u>changes</u> for different materials — this difference is due to their different <u>optical densities</u> (see previous page).

Investigating Light

You've met <u>reflection</u> before — in daily life (every time you look in a mirror for example), and in more physicsy terms on page 130. Well, here's an <u>investigation</u> into reflection. It doesn't get much more fun than this.

How Light **Reflects** Depends on the **Smoothness** of the Surface

This investigation allows you to <u>compare</u> how different surfaces reflect light, as well as confirming the <u>rule</u> for reflection given on page 130.

1) Take a piece of paper and <u>draw a straight line across it</u>. Place an object so one of its sides <u>lines up</u> with this line.

For a refresher on reflection, go back to page 130.

2) Shine a <u>ray</u> of light at the object's surface and <u>trace</u> the <u>incoming</u> and <u>reflected</u> light beams.

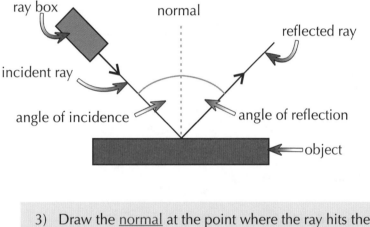

3) Draw the <u>normal</u> at the point where the ray hits the object. Use a protractor to measure the angle of <u>incidence</u> and the <u>angle of reflection</u> and <u>record</u> these values in a table. Also make a note of the <u>width</u> and <u>brightness</u> of the <u>reflected</u> light ray.

4) <u>Repeat</u> this experiment for a range of <u>objects</u>.

You should also see that <u>smooth</u> surfaces like mirrors give <u>clear reflections</u> (the reflected ray is as <u>thin</u> and <u>bright</u> as the <u>incident</u> ray). <u>Rough</u> surfaces like paper cause <u>diffuse</u> reflection (p.130) which causes the reflected beam to be <u>wider and dimmer</u> (or <u>not observable at all</u>). You should also find that:

> The angle of incidence ALWAYS equals the angle of reflection.

 ## Check your results are repeatable...

It's good to <u>repeat</u> this investigation, and the one on the previous page, just in case anything weird happened. You could have set up the experiment <u>wrong</u>, or you might just have read the wrong number off the protractor. Having <u>repeat results</u> will make these odd results easy to spot.

Warm-Up & Exam Questions

Now to check what's actually stuck in your mind over the last eight pages...

Warm-Up Questions

1) Describe the direction of vibrations in a longitudinal wave.
2) Give the formula for calculating the speed of a wave.
3) Outline a method you could use to measure the speed of water waves in a ripple tank.
4) True or false? A wave entering a new medium along the path of the normal won't be refracted.
5) How could you compare the refraction of light through two transparent blocks?

Exam Questions

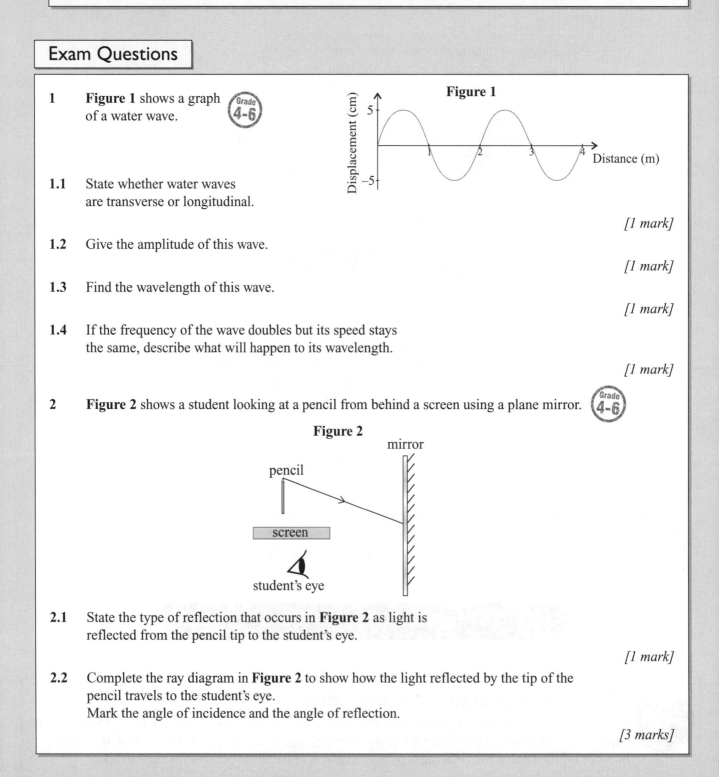

1 **Figure 1** shows a graph of a water wave. **Grade 4-6**

1.1 State whether water waves are transverse or longitudinal.

[1 mark]

1.2 Give the amplitude of this wave.

[1 mark]

1.3 Find the wavelength of this wave.

[1 mark]

1.4 If the frequency of the wave doubles but its speed stays the same, describe what will happen to its wavelength.

[1 mark]

2 **Figure 2** shows a student looking at a pencil from behind a screen using a plane mirror. **Grade 4-6**

2.1 State the type of reflection that occurs in **Figure 2** as light is reflected from the pencil tip to the student's eye.

[1 mark]

2.2 Complete the ray diagram in **Figure 2** to show how the light reflected by the tip of the pencil travels to the student's eye.
Mark the angle of incidence and the angle of reflection.

[3 marks]

Exam Questions

3   **Figure 3** shows a trace representing a sound wave on an
oscilloscope screen. It is a displacement-time graph, which
shows the displacement of a single particle as time passes.

Grade 6-7

Figure 3

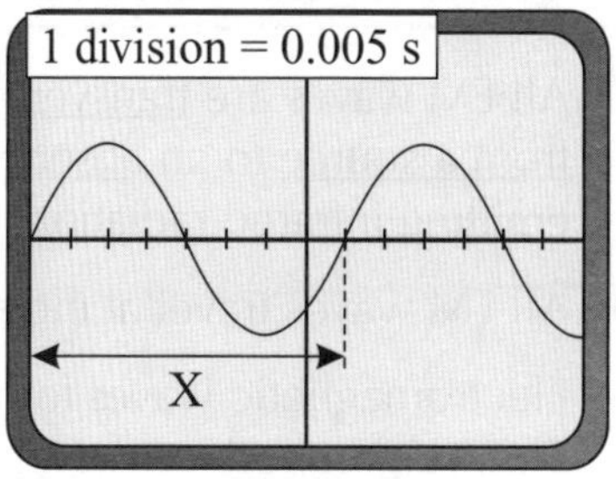

1 division = 0.005 s

3.1   The particle shown by the wave on **Figure 3** is at its
undisturbed (rest) position at T = 0 seconds.
At what time will it next cross the undisturbed position?

[2 marks]

3.2   What quantity does the distance marked X represent?
Tick **one** box.

amplitude ☐   frequency ☐   period ☐

[1 mark]

3.3   Calculate the frequency of the wave.

[4 marks]

4   **Figure 4** shows how an oscilloscope can be used to display sound waves by
connecting microphones to it. Trace 1 shows the sound waves detected by
microphone 1 and trace 2 shows the sound waves detected by microphone 2.

Grade 6-7

Figure 4

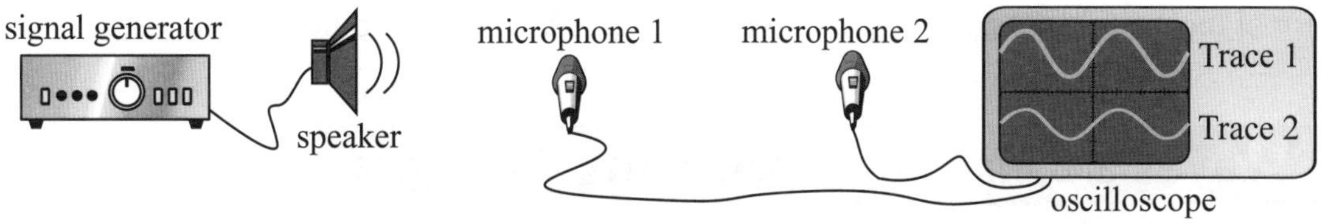

signal generator   microphone 1   microphone 2

speaker

Trace 1

Trace 2

oscilloscope

A student begins with both microphones at equal distances from the speaker and the signal generator
set at a fixed frequency. He gradually moves microphone 2 away from the speaker, which causes
trace 2 to move. He stops moving microphone 2 when both traces line up again as shown in **Figure 4**.
He then measures the distance between the microphones.

4.1   Explain how his measurement could be used to work out the speed of sound in air.

[2 marks]

4.2   With the signal generator set to 50 Hz, the distance between the microphones was measured as 6.8 m.
Calculate the speed of sound in air. Give the correct unit.

[3 marks]

PRACTICAL

5   **Figure 5** shows a ray of red light entering a glass prism.

Grade 7-9

Figure 5

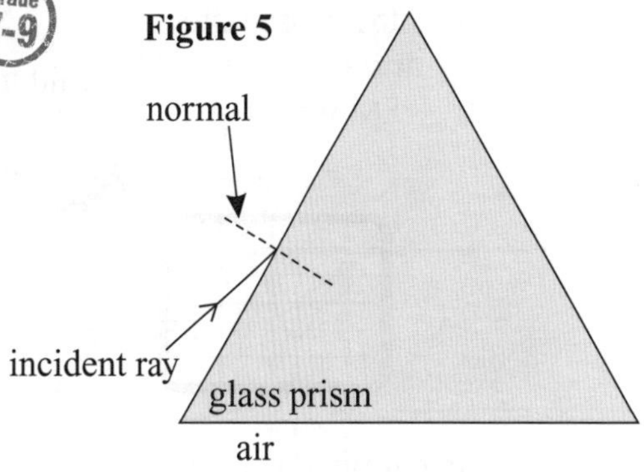

5.1   Complete the diagram to show the ray passing
through the prism and emerging from the other side.
Label the angles of incidence, *I,* and refraction, *R,*
for both boundaries.

[3 marks]

normal

5.2   Describe an experiment that could be carried
out to measure *I* and *R* at both boundaries.

[4 marks]

incident ray

glass prism

air

Electromagnetic Waves and Uses of EM Waves

The differences between types of electromagnetic (EM) waves make them useful to us in different ways.

There's a **Continuous Spectrum** of **EM Waves**

1) All EM waves are transverse waves (p.127) that transfer energy from a source to an absorber. E.g. a hot object transfers energy by emitting infrared radiation, which is absorbed by the surrounding air.

Electromagnetic waves aren't vibrations of particles, they're vibrations of electric and magnetic fields. This means they can travel through a vacuum.

2) All EM waves travel at the same speed through air or a vacuum (space).

3) Electromagnetic waves form a continuous spectrum over a range of frequencies. They're grouped into seven basic types, based on their wavelength and frequency.

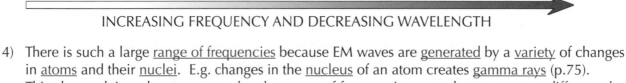

RADIO WAVES	MICRO WAVES	INFRA RED	VISIBLE LIGHT	ULTRA VIOLET	X-RAYS	GAMMA RAYS	Wavelength
$1\ m - 10^4\ m$	$10^{-2}\ m$	$10^{-5}\ m$	$10^{-7}\ m$	$10^{-8}\ m$	$10^{-10}\ m$	$10^{-15}\ m$	

INCREASING FREQUENCY AND DECREASING WAVELENGTH

4) There is such a large range of frequencies because EM waves are generated by a variety of changes in atoms and their nuclei. E.g. changes in the nucleus of an atom creates gamma rays (p.75). This also explains why atoms can absorb a range of frequencies — each one causes a different change.

5) Because of their different properties, different EM waves are used for different purposes.

Radio Waves are Made by **Oscillating Charges**

Head on over to page 52 for more on ac.

1) EM waves are made up of oscillating electric and magnetic fields.

2) Alternating currents (ac) (p.52) are made up of oscillating charges. As the charges oscillate, they produce oscillating electric and magnetic fields, i.e. electromagnetic waves.

3) The frequency of the waves produced will be equal to the frequency of the alternating current.

4) You can produce radio waves using an alternating current in an electrical circuit. The object in which charges (electrons) oscillate to create the radio waves is called a transmitter.

5) When transmitted radio waves reach a receiver, the radio waves are absorbed.

6) The energy transferred by the waves is transferred to the electrons in the material of the receiver.

7) This energy causes the electrons to oscillate and, if the receiver is part of a complete electrical circuit, it generates an alternating current (p.52).

8) This current has the same frequency as the radio waves that generated it.

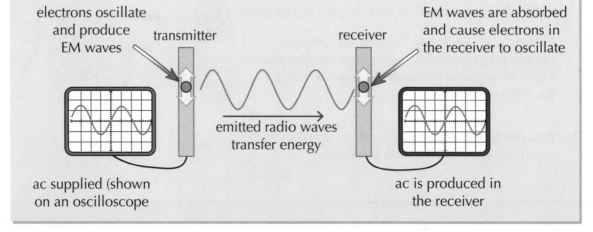

electrons oscillate and produce EM waves · transmitter · receiver · EM waves are absorbed and cause electrons in the receiver to oscillate

emitted radio waves transfer energy

ac supplied (shown on an oscilloscope

ac is produced in the receiver

Uses of EM Waves

Radio waves and microwaves are both types of EM waves, and they're both used for communications. Their exact properties determine which sort of communications they're used for.

Radio Waves are Used Mainly for Communication

1) Radio waves are EM radiation with wavelengths longer than about 10 cm.

2) Long-wave radio waves (wavelengths of 1 – 10 km) can be transmitted from London, say, and received halfway round the world. That's because long wavelengths diffract (bend) around the curved surface of the Earth. Long-wave radio wavelengths can also diffract around hills, into tunnels and all sorts.

3) This makes it possible for radio signals to be received even if the receiver isn't in line of the sight of the transmitter.

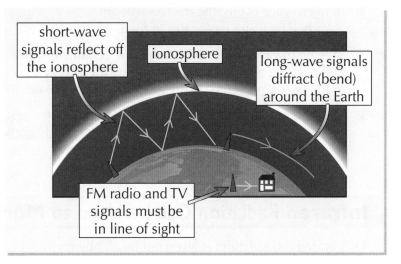

short-wave signals reflect off the ionosphere

ionosphere

long-wave signals diffract (bend) around the Earth

FM radio and TV signals must be in line of sight

4) Short-wave radio signals (wavelengths of about 10 m – 100 m) can, like long-wave, be received at long distances from the transmitter. That's because they are reflected (see p.130) from the ionosphere — an electrically charged layer in the Earth's upper atmosphere.

5) Bluetooth® uses short-wave radio waves to send data over short distances between devices without wires (e.g. wireless headsets so you can use your phone while driving a car).

6) Medium-wave signals (well, the shorter ones) can also reflect from the ionosphere, depending on atmospheric conditions and the time of day.

7) The radio waves used for TV and FM radio transmissions have very short wavelengths. To get reception, you must be in direct sight of the transmitter — the signal doesn't bend or travel far through buildings.

Microwaves are Used by Satellites

Communication to and from satellites (including satellite TV signals and satellite phones) uses microwaves. It's best to use microwaves which can pass easily through the Earth's watery atmosphere.

For satellite TV, the signal from a transmitter is transmitted into space...

... where it's picked up by the satellite receiver dish orbiting thousands of kilometres above the Earth. The satellite transmits the signal back to Earth in a different direction...

... where it's received by a satellite dish on the ground. There is a slight time delay between the signal being sent and received because of the long distance the signal has to travel.

satellite above Earth's atmosphere

microwaves sent back to Earth

microwaves sent to satellite

cloud and water vapour

Uses of EM Waves

Each type of EM wave covers a <u>spectrum</u> of <u>wavelengths</u> and <u>frequencies</u> itself. So the <u>properties</u> of, say, a microwave from one end of the range may <u>differ</u> from those of a microwave from the <u>other end</u> of the range.

Microwave Ovens Also Use Microwaves

1) In <u>microwave ovens</u>, the microwaves are <u>absorbed</u> by <u>water molecules</u> in food.

2) The microwaves penetrate up to a few centimetres into the food before being <u>absorbed</u> and <u>transferring</u> the energy they are carrying to the <u>water molecules</u> in the food, causing the water to <u>heat up</u>.

3) The water molecules then <u>transfer</u> this energy to the rest of the molecules in the food <u>by heating</u> — which <u>quickly cooks</u> the food.

Infrared Radiation Can be Used to Monitor Temperature...

1) <u>Infrared</u> (IR) radiation is <u>given out</u> by all <u>objects</u> — and the <u>hotter</u> the object, the <u>more</u> IR radiation it gives out.

2) <u>Infrared cameras</u> can be used to detect infrared radiation and <u>monitor temperature</u>. The camera detects the IR radiation and turns it into an <u>electrical signal</u>, which is <u>displayed on a screen</u> as a picture. The <u>hotter</u> an object is, the <u>brighter</u> it appears. E.g. <u>energy transfer</u> from a house's <u>thermal energy store</u> can be detected using <u>infrared cameras</u>.

Different <u>colours</u> represent different <u>amounts</u> of <u>IR radiation</u> being detected. Here, the <u>redder</u> the colour, the <u>more</u> infrared radiation is being detected.

...Or Increase it

1) <u>Absorbing</u> IR radiation causes objects to get <u>hotter</u>. <u>Food</u> can be <u>cooked</u> using IR radiation — the <u>temperature</u> of the food increases when it <u>absorbs</u> IR radiation, e.g. from a toaster's heating element.

2) <u>Electric heaters</u> heat a room in the same way. Electric heaters contain a <u>long piece of wire</u> that <u>heats up</u> when a current flows through it. This wire then <u>emits</u> lots of <u>infrared radiation</u> (and a little <u>visible light</u> — the wire <u>glows</u>). The emitted IR radiation is <u>absorbed</u> by objects and the air in the room — energy is transferred <u>by the IR waves</u> to the <u>thermal energy stores</u> of the objects, causing their <u>temperature</u> to <u>increase</u>.

The uses of EM waves depend on their properties...

Differences in wavelength, frequency and energy between types of EM wave give them <u>different properties</u>. For example, some types of EM wave are <u>very harmful</u> (see page 140). Luckily, radio waves are considered <u>safe</u> to beam round the world. IR radiation is generally fairly safe, although too much of it will burn you.

Uses of EM Waves

Here are just a few more uses of EM waves — complete with the all-important <u>reasons</u> why they're used.

Fibre Optic Cables Use Visible Light to Transmit Data

1) <u>Optical fibres</u> are thin <u>glass or plastic fibres</u> that can <u>carry data</u> (e.g. from telephones or computers) over long distances as pulses of <u>visible light</u>.

2) They work because of <u>reflection</u> (p.130). The light rays are <u>bounced back and forth</u> until they reach the end of the fibre.

3) Light is not easily <u>absorbed</u> or <u>scattered</u> as it travels along a fibre.

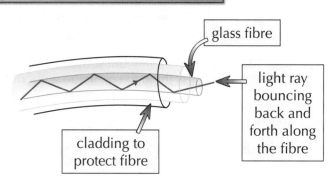

glass fibre

light ray bouncing back and forth along the fibre

cladding to protect fibre

Ultraviolet Radiation Gives You a Suntan

1) <u>Fluorescence</u> is a property of certain chemicals, where <u>ultra-violet</u> (<u>UV</u>) radiation is <u>absorbed</u> and then <u>visible light</u> is <u>emitted</u>. That's why fluorescent colours look so <u>bright</u> — they actually <u>emit light</u>.

2) <u>Fluorescent lights</u> generate <u>UV radiation</u>, which is absorbed and <u>re-emitted as visible light</u> by a layer of a compound called a <u>phosphor</u> on the inside of the bulb. They're <u>energy-efficient</u> (p.28) so they're good to use when light is needed for <u>long periods</u> (like in your <u>classroom</u>).

3) <u>Security pens</u> can be used to <u>mark</u> property with your name (e.g. laptops). Under <u>UV light</u> the ink will <u>glow</u> (fluoresce), but it's <u>invisible</u> otherwise. This can help the police <u>identify</u> your property if it's stolen.

4) <u>Ultraviolet radiation (UV)</u> is produced by the Sun, and exposure to it is what gives people a <u>suntan</u>.

5) When it's <u>not sunny</u>, some people go to <u>tanning salons</u> where <u>UV lamps</u> are used to give them an artificial <u>suntan</u>. However, overexposure to UV radiation can be <u>dangerous</u> (fluorescent lights emit very little UV — they're totally safe).

There's more on the dangers of UV on p.140.

X-rays and Gamma Rays are Used in Medicine

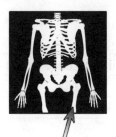

The <u>brighter bits</u> are where <u>fewer X-rays</u> get through. This is a <u>negative image</u>. The plate starts off <u>all white</u>.

There's more on gamma rays on p.75.

1) <u>Radiographers</u> in <u>hospitals</u> take <u>X-ray 'photographs'</u> of people to see if they have any <u>broken bones</u>.

2) X-rays pass <u>easily through flesh</u> but not so easily through <u>denser material</u> like <u>bones</u> or <u>metal</u>. So it's the amount of radiation that's <u>absorbed</u> (or <u>not absorbed</u>) that gives you an X-ray image.

3) <u>Radiographers</u> use <u>X-rays</u> and <u>gamma rays</u> to treat people with <u>cancer</u> (radiotherapy). This is because high doses of these rays <u>kill all living cells</u> — so they are carefully <u>directed</u> towards cancer cells, to avoid killing too many normal, <u>healthy cells</u>.

4) Gamma radiation can also be used as a <u>medical tracer</u> (p.83) — this is where a <u>gamma-emitting source</u> is injected into the patient, and its <u>progress</u> is followed around the body. Gamma radiation is well <u>suited</u> to this because it can <u>pass out</u> through the body to be <u>detected</u>.

5) <u>Both</u> X-rays and gamma rays can be <u>harmful</u> to people (p.140), so radiographers wear <u>lead aprons</u> and stand behind a <u>lead screen</u> or <u>leave the room</u> to keep their exposure to them to a minimum.

Dangers of Electromagnetic Waves

Okay, so you know how <u>useful</u> electromagnetic radiation can be — well, it can also be pretty <u>dangerous</u>.

Some **EM Radiation** Can be **Harmful** to **People**

1) When EM radiation enters <u>living tissue</u>, like <u>you</u>, it's often harmless, but sometimes it creates havoc. The effects of each type of radiation are based on <u>how much energy the wave transfers</u>.

2) <u>Low frequency</u> waves, like <u>radio waves</u>, don't transfer much energy and so mostly <u>pass through soft tissue</u> without being absorbed.

3) <u>High frequency</u> waves like <u>UV</u>, <u>X-rays</u> and <u>gamma rays</u> all transfer <u>lots</u> of energy and so can cause <u>lots of damage</u>.

4) <u>UV radiation</u> damages surface cells, which can lead to <u>sunburn</u> and cause <u>skin</u> to <u>age prematurely</u>. Some more serious effects are <u>blindness</u> and an <u>increased risk of skin cancer</u>.

5) <u>X-rays</u> and <u>gamma rays</u> are types of <u>ionising radiation</u>. (They carry enough energy to <u>knock electrons off of atoms</u>.) This can cause <u>gene mutation or cell destruction</u>, and <u>cancer</u>.

You Can **Measure Risk** Using the **Radiation Dose** in **Sieverts**

1) Whilst UV radiation, X-rays and gamma rays can all be <u>harmful</u>, they are also very <u>useful</u> (see pages 83 and 139). <u>Before</u> any of these types of EM radiation are used, people look at whether the <u>benefits outweigh the health risks</u>.

Radiation doses can be calculated for all types of radiation, not just UV, X-rays and gamma rays.

2) For example, the <u>risk</u> of a person involved in a car accident then developing cancer from having an X-ray photograph taken, is <u>much smaller</u> than the potential health risk of not finding and treating their injuries.

3) <u>Radiation dose</u> (measured in <u>sieverts</u>) is a measure of the <u>risk</u> of harm from the body being exposed to radiation.

4) This is <u>not</u> a measure of the <u>total amount</u> of radiation that has been <u>absorbed</u>.

5) The risk depends on the <u>total amount of radiation</u> absorbed <u>and</u> how <u>harmful</u> the <u>type</u> of radiation is.

6) A sievert is pretty big, so you'll often see doses in <u>millisieverts</u> (mSv), where <u>1000 mSv = 1 Sv</u>.

Risk can be **Different** for **Different Parts** of the **Body**

A CT scan uses <u>X-rays</u> and a <u>computer</u> to build up a picture of the inside of a patient's body. The table shows the <u>radiation dose</u> received by two <u>different</u> <u>parts</u> of a patient's body when having CT scans.

	Radiation dose (mSv)
Head	2.0
Chest	8.0

If a patient has a CT scan on their <u>chest</u>, they are <u>four times more likely</u> to suffer damage to their genes (and their <u>added risk</u> of harm is <u>four times higher</u>) than if they had a <u>head</u> scan.

The risks and benefits of radiation exposure must be balanced...

<u>Ionising radiation</u> can be <u>dangerous</u>, but the risk can be worth taking. From 1920-1970, <u>X-ray machines</u> were installed in shoe shops for use in <u>shoe fittings</u>. But when people realised radiation was harmful, they were phased out. The <u>risks</u> far <u>outweighed</u> the <u>benefits</u> of using X-rays rather than tape measures...

Warm-Up & Exam Questions

There's quite a few different sorts of electromagnetic waves — and you never know which ones might come up in the exams... So check which you're still a bit hazy on with these questions.

Warm-Up Questions

1) Which type of EM wave has the highest frequency?
2) Explain how an alternating current produces radio waves.
3) Explain how microwaves heat food.
4) What are optical fibres used for in phone lines?
5) Why is ionising radiation dangerous?

Exam Questions

1   EM radiation can be harmful but also useful.   (Grade 4-6)

1.1   Give **one** practical use of ultraviolet radiation.

[1 mark]

1.2   Give **one** hazard associated with ultraviolet radiation.

[1 mark]

1.3   Gamma rays can be used in medical tracers to check that your body is working correctly. Explain how medical tracers work.

[3 marks]

1.4   Explain why gamma rays are suitable for use in medical tracers.

[1 mark]

2   **Figure 1** shows a transmitter which is transmitting a communications signal. There is a receiver for the signal inside the house. There is a mountain between the transmitter and the house.   (Grade 6-7)

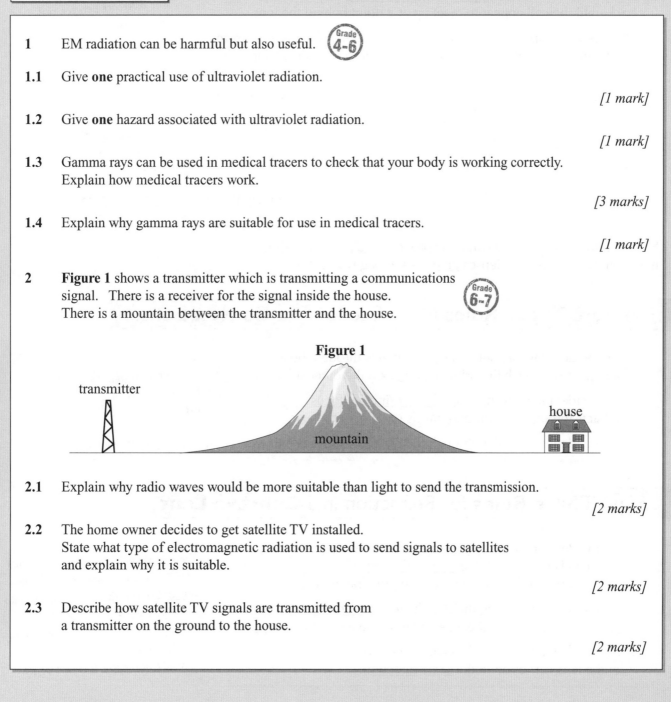

Figure 1

transmitter

house

mountain

2.1   Explain why radio waves would be more suitable than light to send the transmission.

[2 marks]

2.2   The home owner decides to get satellite TV installed. State what type of electromagnetic radiation is used to send signals to satellites and explain why it is suitable.

[2 marks]

2.3   Describe how satellite TV signals are transmitted from a transmitter on the ground to the house.

[2 marks]

Lenses

This bit is about <u>how light acts</u> when it hits a <u>lens</u>. Be ready for lots of diagrams on the next few pages.

Different Lenses Produce Different Kinds of Images

Lenses form images by <u>refracting</u> light (p.131) and changing its direction. There are <u>two main types</u> of lens — <u>convex</u> and <u>concave</u>. They have different shapes and have <u>opposite effects</u> on light rays.

1) A <u>convex</u> lens <u>bulges outwards</u>. It causes rays of <u>light</u> parallel to the axis to be <u>brought together</u> (<u>converge</u>) at the <u>principal focus</u>.

> Convex lenses are also called converging lenses, and concave lenses are also called diverging lenses.

2) A <u>concave</u> lens <u>caves inwards</u>.
   It causes rays of <u>light</u> parallel to axis to <u>spread out</u> (<u>diverge</u>).

3) The <u>axis</u> of a lens is a line passing through the <u>middle</u> of the lens.

4) The <u>principal focus</u> of a <u>convex lens</u> is where rays hitting the lens parallel to the axis all <u>meet</u>.

5) The <u>principal focus</u> of a <u>concave lens</u> is the point where rays hitting the lens parallel to the axis <u>appear</u> to all <u>come from</u> — you can trace them back until they all appear to <u>meet up</u> at a point behind the lens.

> The difference between real and virtual is explained on the next page.

6) There is a principal focus on <u>each side</u> of the lens. The <u>distance</u> from the <u>centre of the lens</u> to the <u>principal focus</u> is called the <u>focal length</u>.

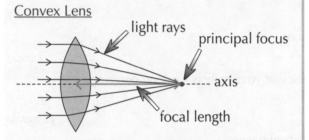

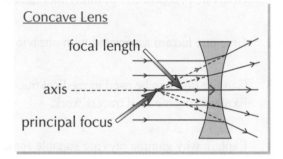

You need to make sure you can draw proper <u>ray diagrams</u> to show how convex and concave lenses <u>differ</u> — see pages 143-144.

There are Three Rules for Refraction in a Convex Lens...

1) An incident ray <u>parallel to the axis</u> refracts through the lens and passes through the <u>principal focus</u> on the other side.

> An example ray diagram for a convex lens is on the next page.

2) An incident ray passing <u>through the principal focus</u> refracts through the lens and travels <u>parallel to the axis</u>.

3) An incident ray passing through the <u>centre</u> of the lens carries on in the <u>same direction</u>.

... And Three Rules for Refraction in a Concave Lens

1) An incident ray <u>parallel to the axis</u> refracts through the lens, and travels in line with the <u>principal focus</u> (so it appears to have come from the principal focus).

> The <u>neat thing</u> about these rules is that they allow you to draw ray diagrams <u>without</u> bending the rays as they go into the lens <u>and</u> as they leave the lens. You can draw the diagrams as if each ray only changes direction <u>once</u>, in the <u>middle of the lens</u> (see next page).

2) An incident ray passing through the lens <u>towards the principal focus</u> refracts through the lens and travels <u>parallel to the axis</u>.

3) An incident ray passing through the <u>centre</u> of the lens carries on in the <u>same direction</u>.

Images and Ray Diagrams

This page is about <u>real</u> and <u>virtual images</u> and how to draw a <u>ray diagram</u> for a <u>convex</u> lens.

Lenses can Produce **Real** and **Virtual** Images

1) A <u>real image</u> is where the <u>light from an object</u> comes together to form an <u>image on a 'screen'</u> — like the image formed on an eye's <u>retina</u> (the 'screen' at the back of an <u>eye</u>).

2) A <u>virtual image</u> is when the rays are diverging, so the light from the object <u>appears</u> to be coming from a completely <u>different place</u>.

3) When you look in a <u>mirror</u> you see a <u>virtual image</u> of your face — because the <u>object</u> (your face) <u>appears</u> to be <u>behind the mirror</u>.

4) You can get a virtual image when looking at an object through a <u>magnifying lens</u> (see p.144) — the virtual image looks <u>bigger</u> than the object <u>actually</u> is.

5) To describe an image properly, you need to say <u>3 things</u>:

- <u>How big it is</u> compared to the object.
- Whether it's <u>upright or inverted</u> (upside down) relative to the object.
- Whether it's <u>real or virtual</u>.

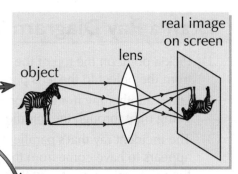

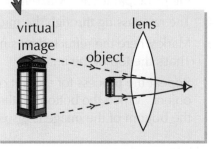

Draw a **Ray Diagram** for an **Image** Through a **Convex Lens**

1) Pick a point on the <u>top</u> of the object. Draw a ray going from the object to the lens <u>parallel</u> to the axis of the lens.

2) Draw another ray from the <u>top</u> of the object going right through the <u>middle</u> of the lens.

3) The incident ray that's <u>parallel</u> to the axis is <u>refracted</u> through the <u>principal focus</u> (F) on the <u>other side</u> of the lens. Draw a <u>refracted ray</u> passing through the <u>principal focus</u>.

4) The ray passing through the <u>middle</u> of the lens doesn't bend.

5) Mark where the rays <u>meet</u>. That's the <u>top of the image</u>.

6) Repeat the process for a point on the bottom of the object. When the bottom of the object is on the axis, the bottom of the image is <u>also</u> on the axis.

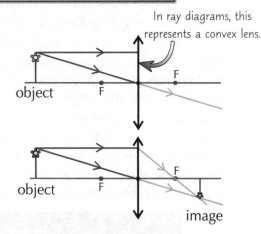

In ray diagrams, this represents a convex lens.

If you really want to draw a third incident ray passing through the principal focus on the way to the lens, you can (refract it so that it goes parallel to the axis).

Distance from the Lens Affects the **Image**

1) An object <u>at 2F</u> will produce a <u>real, inverted</u> image the <u>same size</u> as the object, and <u>at 2F</u>.

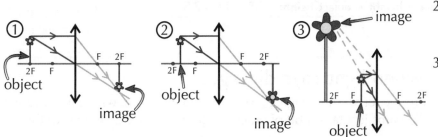

2) <u>Between F and 2F</u> it'll make a <u>real, inverted</u> image <u>bigger</u> than the object, and <u>beyond 2F</u>.

3) An object <u>nearer than F</u> will make a <u>virtual</u> image that is <u>upright</u>, <u>bigger</u> than the object and on the <u>same side</u> of the lens.

Concave Lenses and Magnification

Now for <u>concave lenses</u> and a handy equation — the <u>magnification formula</u>.

Draw a **Ray Diagram** for an **Image** Through a **Concave Lens**

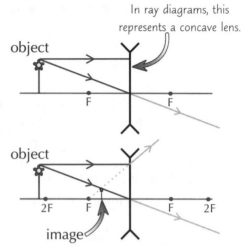

In ray diagrams, this represents a concave lens.

1) Pick a point on the <u>top</u> of the object. Draw a ray going from the object to the lens <u>parallel</u> to the axis of the lens.

2) Draw another ray from the <u>top</u> of the object going right through the <u>middle</u> of the lens.

3) The incident ray that's <u>parallel</u> to the axis is <u>refracted</u> so it appears to have come from the <u>principal focus</u>. Draw a <u>ray</u> from the principal focus. Make it <u>dotted</u> before it reaches the lens.

4) The ray passing through the <u>middle</u> of the lens doesn't bend.

5) Mark where the refracted rays <u>meet</u>. That's the top of the image.

6) Repeat the process for a point on the bottom of the object. When the bottom of the object is on the <u>axis</u>, the bottom of the image is <u>also</u> on the axis.

Unlike a convex lens, a concave lens always produces a <u>virtual image</u>. The image is <u>upright</u>, <u>smaller</u> than the object and on the <u>same side of the lens as the object</u> — <u>no matter where the object is</u>.

Magnifying Glasses Use Convex Lenses

Magnifying glasses work by creating a <u>magnified virtual image</u> (see previous page).

1) The object being magnified must be closer to the lens than the <u>focal length</u>.

2) Since the image produced is a <u>virtual image</u>, the light rays don't <u>actually</u> come from the place where the image appears to be.

3) Remember "you <u>can't</u> project a virtual image onto a screen" — that's a <u>useful phrase</u> to use in the exam if they ask you about virtual images.

4) You can use the <u>magnification formula</u> to work out the magnification produced by a <u>lens</u> at a given distance:

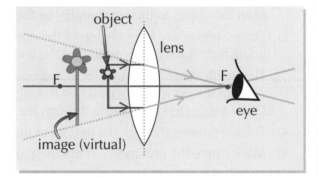

$$\text{magnification} = \frac{\text{image height}}{\text{object height}}$$

Magnification is a ratio, so it doesn't have any units. This means so long as the units are the same, you can measure the heights in whatever units you like.

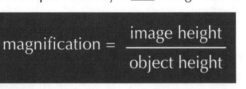

A coin with diameter 14 mm is placed behind a magnifying lens. The virtual image produced has a diameter of 35 mm. What is the magnification of the lens at this distance?

magnification = image height ÷ object height = 35 ÷ 14 = 2.5

You can also find the magnification by dividing the distance between the image and the lens by the distance between the object and the lens.

You get a virtual image when light rays diverge...

... like when they come through a <u>concave</u> lens. The image <u>appears</u> to be at the point where the diverging rays <u>seem</u> to have come from. Ray diagrams just take a bit of practice. It's the only way to master them.

Visible Light

We see light all of the time. But it's a bit more complicated than you might have thought.

Visible Light is Made Up of a Range of Colours

1) As you saw on page 136, EM waves cover a very large spectrum. We can only see a tiny part of this — the visible light spectrum. This is a range of wavelengths that we perceive as different colours.

2) Each colour has its own narrow range of wavelengths (and frequencies) ranging from violets down at 400 nm up to reds at 700 nm.

3) Different mixtures of these colours allow us to see even more shades, e.g. red and blue wavelengths seen together appear purple. When all of these different colours are put together, it creates white light.

Colour and Transparency Depend on Absorbed Wavelengths

1) Different objects absorb, transmit and reflect different wavelengths of light in different ways (p.130).

2) Opaque objects are objects that do not transmit light. When visible light waves hit them, they absorb some wavelengths of light and reflect others.

3) The colour of an opaque object depends on which wavelengths of light are most strongly reflected. E.g. a red apple appears to be red because the wavelengths corresponding to the red part of the visible spectrum are most strongly reflected. The other wavelengths of light are absorbed.

The colours our brains 'see' depends on other things like brightness and how much of each wavelength enters our eyes.

White light (a combination of colours) hits the apple...

...red light is reflected — all other colours are absorbed. The apple looks red.

4) For opaque objects that aren't a primary colour, they may be reflecting either the wavelengths of light corresponding to that colour OR the wavelengths of the primary colours that can mix together to make that colour. So a banana may look yellow because it's reflecting yellow light OR because it's reflecting both red and green light.

5) White objects reflect all of the wavelengths of visible light equally.

6) Black objects absorb all wavelengths of visible light. Your eyes see black as the lack of any visible light (i.e. the lack of any colour).

7) Transparent (see-through) and translucent (partially see-through) objects transmit light, i.e. not all light that hits the surface of the object is absorbed or reflected — some can pass through.

8) Some wavelengths of light may be absorbed or reflected by transparent and translucent objects. A transparent or translucent object's colour is related to the wavelengths of light transmitted and reflected by it.

We see the colour of light that objects reflect into our eyes...

So the only thing that makes red stuff red is the fact that it reflects the 'red' wavelengths of light and absorbs the rest. So what colour is it in the dark when there's no light to reflect? The mind boggles.

Filters

Colour filters are <u>transparent objects</u> that absorb most wavelengths of light and just let some through. E.g. red cellophane — everything viewed though it appears to be either a shade of red or black. Here's why...

Colour Filters Only **Let Through Particular Wavelengths**

1) Colour filters are used to <u>filter out</u> different <u>wavelengths</u> of light, so that only certain colours (wavelengths) are <u>transmitted</u> — the rest are <u>absorbed</u>.

2) A <u>primary colour filter</u> only <u>transmits</u> that <u>colour</u>, e.g. if <u>white light</u> is shone at a <u>blue</u> colour filter, <u>only</u> blue light will be let through. The rest of the light will be <u>absorbed</u>.

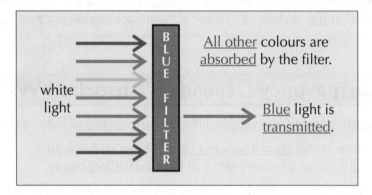

<u>All other</u> colours are <u>absorbed</u> by the filter.

white light

<u>Blue</u> light is <u>transmitted</u>.

3) If you look at a <u>blue object</u> through a blue <u>colour filter</u>, it would still look <u>blue</u>. Blue light is <u>reflected</u> from the object's surface and is <u>transmitted</u> by the filter.

4) However, if the object was e.g. <u>red</u> (or any colour <u>not made from blue</u> <u>light</u>), the object would appear <u>black</u> when viewed through a blue filter. <u>All</u> of the light <u>reflected</u> by the object will be <u>absorbed</u> by the filter.

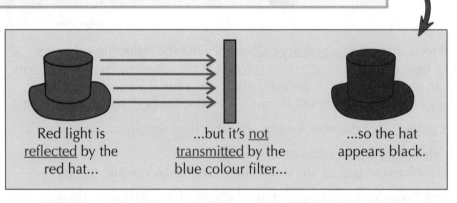

Red light is <u>reflected</u> by the red hat...

...but it's <u>not</u> <u>transmitted</u> by the blue colour filter...

...so the hat appears black.

5) <u>Filters</u> that <u>aren't</u> for <u>primary</u> colours let through <u>both</u> the <u>wavelengths</u> of light for that <u>colour</u> AND the wavelengths of the <u>primary</u> colours that can be added together to make that colour.

Colour filters only let their own colour of light through...

Hopefully you now know enough about <u>absorption</u> and <u>reflection</u> that you're feeling pretty confident about how colour filters work. Once you absorb the basic facts, the stuff on this page and on the previous page is pretty easy — red objects reflect red light and red filters let red light through. Simple.

Warm-Up & Exam Questions

It's best to make sure this stuff's crystal clear now, or you'll regret it when you get to the exam... So here's lots of questions on lenses, as well as on colour and filters. Go on, answer them, you know you want to.

Warm-Up Questions

1) Name the two main types of lens.
2) Of the two types of lens, which always creates a virtual image and which can create both real and virtual images?
3) Explain why a red object appears red.
4) Why does a green object viewed through a red filter appear black?

Exam Questions

1 This question is about different types of lens. **Grade 6-7**

1.1 Ed is trying to start a campfire by focussing sunlight through his spectacle lens onto the firewood.
The lens is concave.
Explain why he cannot focus the sunlight onto the wood using this lens.

[2 marks]

1.2 Ed finds a slug and uses a magnifying glass to look at it.
Complete **Figure 1** to show where the image of the slug is formed.

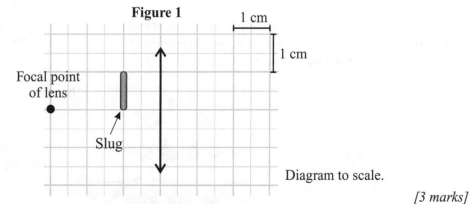

Figure 1

1 cm

1 cm

Focal point
of lens

Slug

Diagram to scale.

[3 marks]

1.3 Calculate the magnification of the lens for the slug at this distance.
Use the correct equation from the Physics Equation Sheet on the inside back cover.

[2 marks]

2 **Figure 2** shows the distribution of wavelengths of visible light emitted by a particular flat screen, compared to that of normal daylight. **Grade 7-9**

Figure 2

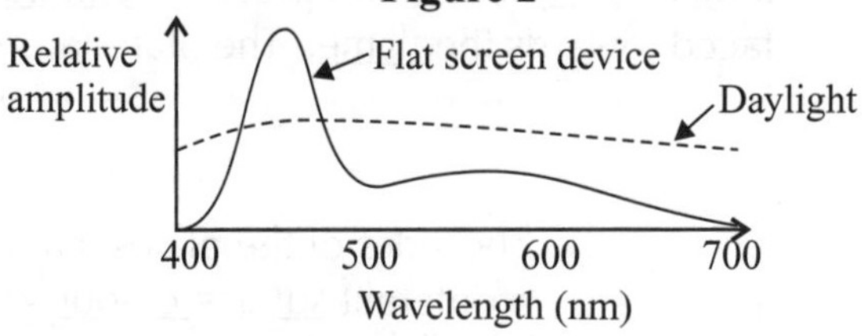

Relative amplitude

Flat screen device

Daylight

400   500   600   700
Wavelength (nm)

2.1 Describe how light from the flat screen device differs from daylight.

[3 marks]

2.2 Exposure to blue light, which has wavelengths below 500 nm, has been linked to eye problems.
Suggest how people could protect themselves from these problems while using the device.
Explain your answer.

[2 marks]

Infrared Radiation and Temperature

Infrared radiation is what you feel as <u>heat</u>. It's easy to think that only really hot objects, like the glowing bars on an electric fire, give out infrared radiation, but that's <u>not</u> at all true as you'll soon discover...

Every Object Absorbs and Emits Infrared Radiation

<u>All objects</u> are <u>continually emitting</u> and <u>absorbing infrared</u> (IR) radiation.

1) Infrared radiation is emitted from the <u>surface</u> of an object.

2) The <u>hotter</u> an object is, the <u>more</u> infrared radiation it radiates in a given time.

3) An object that's <u>hotter</u> than its surroundings <u>emits more</u> <u>IR radiation</u> than it <u>absorbs</u> as it <u>cools down</u> (e.g. a cup of tea left on a table). And an object that's <u>cooler</u> than its surroundings <u>absorbs</u> more IR radiation than it <u>emits</u> as it <u>warms up</u> (e.g. a cold glass of water on a sunny day).

4) Objects at a <u>constant temperature</u> emit infrared radiation at the <u>same rate</u> that they are <u>absorbing it</u>.

5) <u>Some colours</u> and <u>surfaces absorb</u> and <u>emit</u> radiation better than others. For example, a <u>black</u> surface is <u>better</u> at absorbing and emitting radiation than a <u>white</u> one, and a <u>matt</u> surface is <u>better</u> at absorbing and emitting radiation than a <u>shiny</u> one.

The hot chocolate (and the mug) is warmer than the air around it, so it gives out more IR radiation than it absorbs, which cools it down.

You Can Investigate Absorption with the Melting Wax Trick

The amount of infrared radiation <u>absorbed</u> by different materials also depends on the <u>material</u>. You can do an experiment to show this, using a <u>Bunsen burner</u> and some <u>candle wax</u>.

PRACTICAL

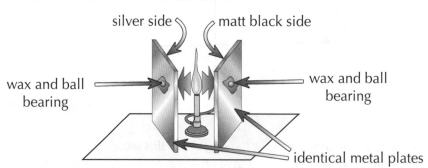

silver side   matt black side

wax and ball bearing

wax and ball bearing

identical metal plates

1) Set up the equipment as shown above. Two <u>ball bearings</u> are each stuck to <u>one side</u> of a <u>metal plate</u> with solid pieces of <u>candle wax</u>. The other sides of these plates are then faced towards the <u>flame</u>. The plates are placed the <u>same distance</u> away from the flame.

2) The sides of the plates that are facing towards the flame each have a different <u>surface colour</u> — one is <u>matt black</u> and the other is <u>silver</u>.

3) The ball bearing on the black plate will <u>fall first</u> as the black surface <u>absorbs more</u> infrared radiation — <u>transferring</u> more energy to the <u>thermal energy store</u> of the wax. This means the wax on the <u>black</u> plate melts <u>before</u> the wax on the <u>silver</u> plate.

Investigating Emission PRACTICAL

Time for another <u>Required Practical</u>. In this one, you'll meet a fun, new piece of kit called a <u>Leslie Cube</u>. Read on to find out more about how you can use this equipment to investigate <u>infrared radiation emissions</u>.

You Can Investigate **Emission** With a **Leslie Cube**

A <u>Leslie cube</u> is a <u>hollow</u>, <u>watertight</u>, metal cube made of e.g. aluminium, whose four <u>vertical faces</u> have <u>different surfaces</u> (for example, matt black paint, matt white paint, shiny metal and dull metal). You can use them to <u>investigate IR radiation emitted</u> by different surfaces:

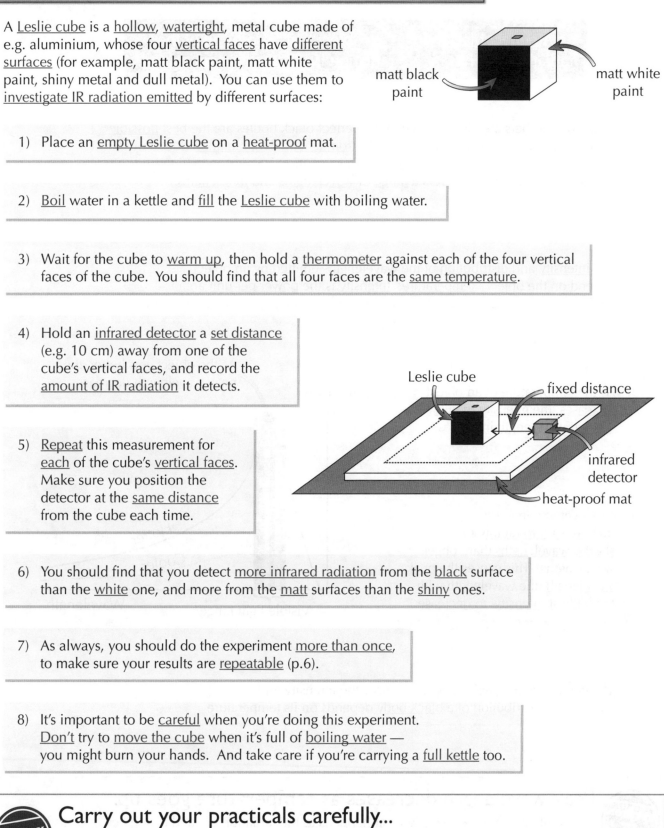

matt black paint

matt white paint

1) Place an <u>empty Leslie cube</u> on a <u>heat-proof</u> mat.

2) <u>Boil</u> water in a kettle and <u>fill</u> the <u>Leslie cube</u> with boiling water.

3) Wait for the cube to <u>warm up</u>, then hold a <u>thermometer</u> against each of the four vertical faces of the cube. You should find that all four faces are the <u>same temperature</u>.

4) Hold an <u>infrared detector</u> a <u>set distance</u> (e.g. 10 cm) away from one of the cube's vertical faces, and record the <u>amount of IR radiation</u> it detects.

Leslie cube

fixed distance

infrared detector

heat-proof mat

5) <u>Repeat</u> this measurement for <u>each</u> of the cube's <u>vertical faces</u>. Make sure you position the detector at the <u>same distance</u> from the cube each time.

6) You should find that you detect <u>more infrared radiation</u> from the <u>black</u> surface than the <u>white</u> one, and more from the <u>matt</u> surfaces than the <u>shiny</u> ones.

7) As always, you should do the experiment <u>more than once</u>, to make sure your results are <u>repeatable</u> (p.6).

8) It's important to be <u>careful</u> when you're doing this experiment. <u>Don't</u> try to <u>move the cube</u> when it's full of <u>boiling water</u> — you might burn your hands. And take care if you're carrying a <u>full kettle</u> too.

PRACTICAL TIP

Carry out your practicals carefully...

And that means both being careful when <u>collecting data</u>, and careful when dealing with potential <u>hazards</u>. Watch out when you're pouring or carrying <u>boiling water</u>, and make sure any water or equipment has <u>cooled down</u> enough before you start handling it after your experiment is done.

Black Body Radiation

Black body radiation sounds complicated, and I'm afraid it is a bit, but just take your time and you'll get it.

Black Bodies are the Ultimate Emitters

A perfect black body is an object that absorbs all of the
radiation that hits it. No radiation is reflected or transmitted.

1) As good absorbers are also good emitters, perfect black bodies are the best possible emitters of radiation. (All objects emit electromagnetic (EM) radiation due to the energy in their thermal energy stores. This radiation isn't just in the infrared part of the spectrum — it covers a range of wavelengths and frequencies.)

2) The intensity and distribution of the wavelengths emitted by an object depend on the object's temperature. Intensity is the power per unit area, i.e. how much energy is transferred to a given area in a certain amount of time.

3) As the temperature of an object increases, the intensity of every emitted wavelength increases.

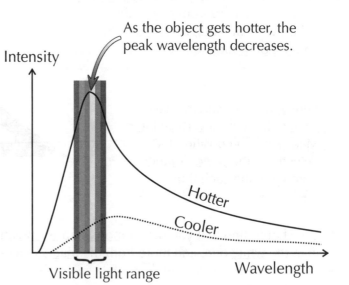

As the object gets hotter, the peak wavelength decreases.

4) However, the intensity increases more rapidly for shorter wavelengths than longer wavelengths. This causes the peak wavelength (the wavelength with the highest intensity) to decrease.

5) The curves on the graph above show how the intensity and wavelength distribution of a black body depends on its temperature.

Peak wavelength decreases as temperature goes up...

REVISION TIP

The graph above shows exactly what happens to the radiation emitted by a black body as it's heated, but it's a good enough model for most objects — and it's also the key to understanding this rather tricky stuff. So make sure you can sketch it and label it from memory — then practice explaining it to yourself. Otherwise you could miss out on some important marks in the exam...

Earth and Radiation

It's lucky for us that the <u>Sun</u> is close by (well, about 150 million km away) and transfers energy to us by <u>EM radiation</u>. It's also lucky for us that a lot of this energy is <u>reflected</u> or <u>emitted</u> back into space.

Radiation Affects the Earth's Temperature

The overall temperature of the Earth depends on the amount of IR radiation it <u>reflects</u>, <u>absorbs</u> and <u>emits</u>.

1) <u>During the day</u>, <u>lots</u> of radiation (like light) is transferred to the Earth from the Sun and <u>absorbed</u>. This causes an <u>increase</u> in <u>local</u> temperature.

2) At <u>night</u>, less radiation is being <u>absorbed</u> than is being <u>emitted</u>, causing a <u>decrease</u> in the <u>local</u> temperature.

3) <u>Overall</u>, the <u>temperature</u> of the Earth stays <u>fairly constant</u>. You can show the flow of <u>radiation</u> for the Earth on a handy <u>diagram</u>.

Some radiation is reflected by the atmosphere, clouds and the Earth's surface.

Some radiation is emitted by the atmosphere.

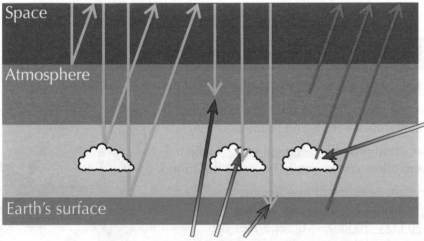

Some of the radiation emitted by the surface is reflected or absorbed (and later emitted) by the clouds.

Some radiation is absorbed by the atmosphere, clouds and the Earth's surface.

4) <u>Changes</u> to the atmosphere can cause a change to the Earth's <u>overall temperature</u>. If the atmosphere starts to <u>absorb</u> more radiation without <u>emitting the same amount</u>, the <u>overall temperature</u> will rise until absorption and emission are <u>equal</u> again.

Absorbed radiation means a rise in temperature...

Greenhouse gases, such as carbon dioxide, are <u>good absorbers</u> of radiation. That's why adding more of them to the atmosphere causes the Earth's atmosphere to <u>warm up</u> as more radiation is absorbed by the atmosphere and less is emitted back into space. This is the mechanism behind <u>global warming</u>.

Sound Waves

Time to learn how we <u>hear</u> things. Don't panic — you won't be tested on each individual part of the <u>ear</u>.

Sound Travels as a Wave

1) <u>Sound waves</u> are caused by <u>vibrating objects</u>. These vibrations are passed through the surrounding medium as a series of <u>compressions</u> and <u>rarefactions</u> (sound is a type of <u>longitudinal wave</u> — page 127).

2) Sound generally travels <u>faster in solids</u> than in liquids, and faster in liquids than in gases.

3) When a sound wave travels <u>through a solid</u> it does so by causing the <u>particles</u> in the solid to <u>vibrate</u>.

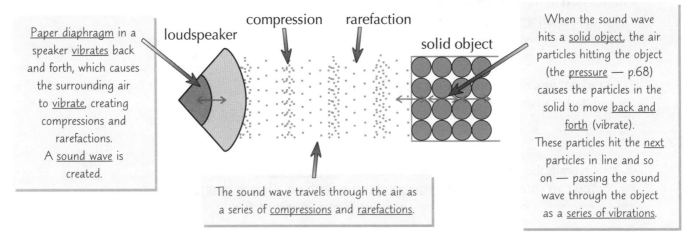

<u>Paper diaphragm</u> in a speaker <u>vibrates</u> back and forth, which causes the surrounding air to <u>vibrate</u>, creating compressions and rarefactions. A <u>sound wave</u> is created.

loudspeaker   compression   rarefaction   solid object

The sound wave travels through the air as a series of <u>compressions</u> and <u>rarefactions</u>.

When the sound wave hits a <u>solid object</u>, the air particles hitting the object (the <u>pressure</u> — p.68) causes the particles in the solid to move <u>back and forth</u> (vibrate). These particles hit the <u>next</u> particles in line and so on — passing the sound wave through the object as a <u>series of vibrations</u>.

4) Sound can't travel in <u>space</u>, because it's mostly a <u>vacuum</u> (there are no particles to move or vibrate).

5) Sometimes the sound wave will eventually travel into someone's <u>ear</u> and reach their <u>ear drum</u> at which point they might <u>hear the sound</u> — more on this below.

You Hear Sound When Your Ear Drum Vibrates

1) Sound waves that reach your <u>ear drum</u> can cause it to <u>vibrate</u>.

2) These <u>vibrations</u> are passed on to <u>tiny bones</u> in your ear called ossicles, through the semicircular canals and to the cochlea.

3) The <u>cochlea</u> turns these vibrations into <u>electrical signals</u> which get sent to your brain and allow you to <u>sense</u> (i.e. <u>hear</u>) the sound.

4) <u>Different materials</u> can convert different <u>frequencies</u> of sound waves into <u>vibrations</u>. E.g. <u>humans</u> can hear sound in the range of <u>20 Hz – 20 kHz</u>. Microphones can pick up sound waves <u>outside</u> of this range, but if you tried to listen to this sound, you probably <u>wouldn't hear anything</u>.

5) Human hearing is limited by the <u>size</u> and <u>shape</u> of our <u>ear drum</u>, as well as the <u>structure</u> of all the parts within the ear that <u>vibrate</u> to transfer the energy from the sound wave.

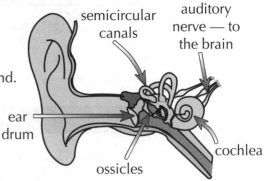

semicircular canals   auditory nerve — to the brain

ear drum

ossicles   cochlea

Microphones work in a similar way. Sound waves cause a diaphragm to vibrate and this movement is transferred into an electrical signal.

Sound Waves Can Reflect and Refract

1) Sound waves will be <u>reflected</u> by <u>hard flat surfaces</u>. <u>Echoes</u> are just reflected sound waves.

2) <u>Sound waves</u> will also <u>refract</u> as they enter <u>different media</u>. As they enter <u>denser material</u>, they <u>speed up</u>. This is because when a wave travels into a different medium, its <u>wavelength changes</u> but its <u>frequency remains the same</u> so its <u>speed</u> must also <u>change</u> (p.127)

However, since sound waves are always spreading out so much, the change in direction is hard to spot under normal circumstances.

Ultrasound

Can you hear <u>that</u>? If not, '<u>that</u>' could be <u>ultrasound</u> — a handy wave used for <u>seeing hidden objects</u>.

Ultrasound is Sound with **Frequencies Higher** Than **20 000 Hz**

1) Electrical devices can be made which produce <u>electrical oscillations</u> of <u>any frequency</u>.

2) These can easily be converted into <u>mechanical vibrations</u> to produce <u>sound</u> waves <u>beyond the range of human hearing</u> (i.e. frequencies above 20 000 Hz).

3) This is called <u>ultrasound</u> and it pops up all over the place.

Ultrasound Waves Get **Partially Reflected** at **Boundaries**

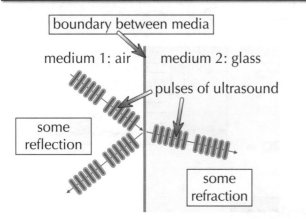

boundary between media

medium 1: air   medium 2: glass

pulses of ultrasound

some reflection

some refraction

1) When a wave passes from one medium into another, <u>some</u> of the wave is <u>reflected</u> off the boundary between the two media, and some is transmitted (and refracted). This is <u>partial reflection</u>.

2) What this means is that you can point a pulse of ultrasound at an object, and wherever there are <u>boundaries</u> between one substance and another, some of the ultrasound gets <u>reflected back</u>.

3) The time it takes for the reflections to reach a <u>detector</u> can be used to measure <u>how far away</u> the boundary is.

Ultrasound has **Medical** and **Industrial Uses**

1) <u>Ultrasound waves</u> can pass through the body, but whenever they reach a boundary between <u>two different media</u> (like fluid in the womb and the skin of the foetus) some of the wave is <u>reflected back</u> and <u>detected</u>.

2) The exact <u>timing and distribution</u> of these <u>echoes</u> are processed by a computer to produce a <u>video image</u> of the foetus.

3) No one knows for sure if ultrasound is safe in all cases but <u>X-rays</u> would definitely be dangerous.

ultrasound transmitter/receiver

partial reflection

foetus

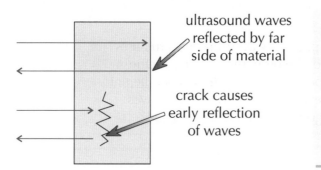

ultrasound waves reflected by far side of material

crack causes early reflection of waves

4) Ultrasound can also be used to find <u>flaws</u> in objects such as <u>pipes</u> or <u>materials</u> such as wood or metal.

5) Ultrasound waves entering a material will usually be <u>reflected</u> by the <u>far side</u> of the material.

6) If there is a flaw such as a <u>crack</u> inside the object, the wave will be <u>reflected sooner</u>.

Learn how ultrasound is used in medicine and industry...

As well as scanning foetuses, ultrasound can be used to image other parts of the body without the <u>risks</u> associated with using radiation (see page 140 for more on the dangers of radiation).

Exploring Structures Using Waves

Ultrasound is really useful. On the previous page, you saw just one of its <u>medical</u> uses.
Now for some of its <u>other</u> uses — but this time, out at <u>sea</u>.

Ultrasound is Used in Underwater Exploration

<u>Echo sounding</u> uses high frequency <u>sound waves</u> (including <u>ultrasound</u>).
It's used by boats and submarines to find out the <u>depth of the water</u> they
are in or to <u>locate</u> objects in <u>deep water</u>.

EXAMPLE:

**A pulse of ultrasound takes 4.5 seconds to travel
from a submarine to the seabed and back again.
If the speed of sound in seawater is 1520 m/s,
how far away is the submarine from the seabed?**

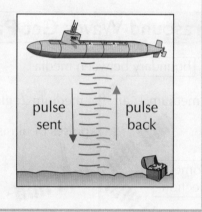

pulse sent | pulse back

1) The formula is of course
   <u>distance = speed × time</u>.

   $s = vt$
   $= 1520 × 4.5$
   $= 6840$

2) But this is a <u>reflection question</u>,
   so don't forget the factor of 2.

3) The 4.5 s is for there and back,
   so <u>halve the distance</u>.

   $6840 ÷ 2 = 3420$ m

Waves Can Be Used to Detect and Explore

1) Waves have <u>different properties</u> (e.g. speed)
   depending on the <u>material</u> they're travelling through.

2) When a wave arrives at a <u>boundary</u> between
   materials, a number of things can happen.

3) It can be <u>completely reflected</u> or <u>partially reflected</u> (like in
   <u>ultrasound</u> imaging, see previous page). The wave may continue
   travelling in the same direction but at a <u>different speed</u>, or it may be
   <u>refracted</u> (p.131) or <u>absorbed</u> (like <u>S-waves</u> — see the next page).

4) Studying the properties and paths of <u>waves</u> through structures
   can give you clues to some of the properties of the structure
   that you can't <u>see</u> by eye. You can do this with lots of <u>different</u>
   <u>waves</u> — ultrasound (see previous page) and seismic waves
   (see the next page) are two good, well-known <u>examples</u>.

Make sure you can use $s = vt$ to work out distances...

You should be familiar with the equation linking <u>distance</u>, <u>speed</u> and <u>time</u> from page 104.
Remember, if you're given the time taken for an <u>ultrasound pulse</u> to travel from an emitter to a
boundary and back, you must <u>divide your answer by two</u> to find just the distance to the boundary.

Seismic Waves

All seismic waves are <u>not</u> the same. There are <u>surface waves</u> which travel along the Earth's surface.
There are also <u>body waves</u> which travel through the Earth — it's these you need to know about.

Earthquakes and Explosions Cause Seismic Waves

1) When there's an <u>earthquake</u> somewhere, it produces <u>seismic waves</u> which travel out through the Earth.
We <u>detect</u> these waves all over the surface of the planet using <u>seismometers</u>.

2) <u>Seismologists</u> work out the <u>time</u> it takes for the shock waves to reach each seismometer.
They also note which parts of the Earth <u>don't receive the shock waves</u> at all.

3) When <u>seismic waves</u> reach a <u>boundary</u> between different layers of <u>material</u> (which all have different
<u>properties</u>, like density) inside the Earth, some waves will be <u>absorbed</u> and some will be <u>refracted</u>.

4) Most of the time, if the waves are <u>refracted</u> they change speed <u>gradually</u>, resulting in a <u>curved path</u>.
But when the properties change <u>suddenly</u>, the wave speed changes abruptly, and the path has a <u>kink</u>.

Seismic Waves Provide Evidence for the Earth's Structure

1) By observing how seismic waves are <u>absorbed</u> and
<u>refracted</u>, scientists have been able to work out <u>where</u>
the properties of the Earth change <u>dramatically</u>.

2) Our current understanding of the <u>internal
structure</u> of the Earth and the <u>size</u> of the
<u>Earth's core</u> is based on these <u>observations</u>.

3) There are <u>two different types</u> of seismic waves you
need to learn — <u>P waves</u> and <u>S waves</u> (see below).

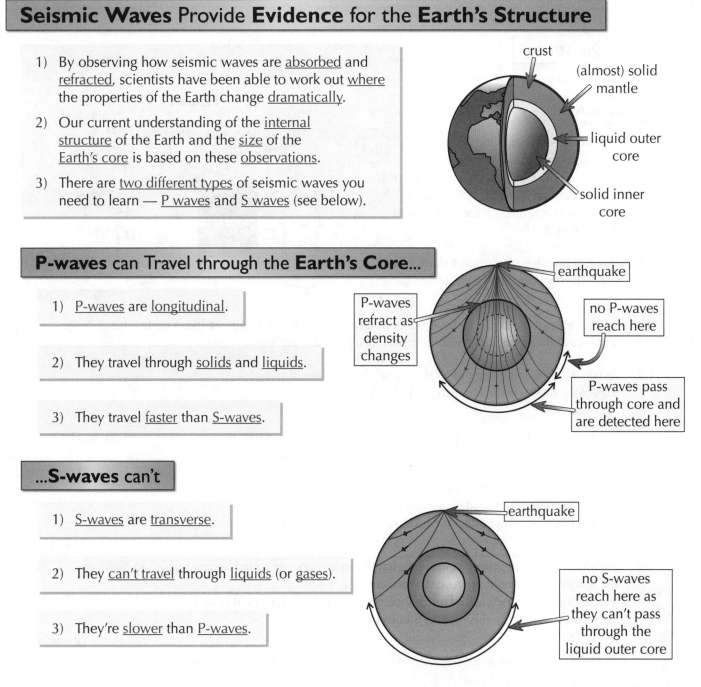

P-waves can Travel through the Earth's Core...

1) <u>P-waves</u> are <u>longitudinal</u>.

2) They travel through <u>solids</u> and <u>liquids</u>.

3) They travel <u>faster</u> than <u>S-waves</u>.

...S-waves can't

1) <u>S-waves</u> are <u>transverse</u>.

2) They <u>can't travel</u> through <u>liquids</u> (or <u>gases</u>).

3) They're <u>slower</u> than <u>P-waves</u>.

Warm-Up & Exam Questions

This is the last set of questions in the waves topic. Phew, I hear you say. There's some tricky stuff on these pages, such as black body radiation. But plough through them carefully and you'll get there.

Warm-Up Questions

1) If an object is hotter than its surroundings, does it emit more or less IR radiation than it absorbs?
2) What specialised piece of apparatus is used to investigate IR radiation emissions from surfaces?
3) In terms of radiation absorbed and emitted, explain why local temperatures drop at night.
4) Give the definition of a perfect black body.
5) What sort of wave is a sound wave — transverse or longitudinal?
6) What does it mean if a wave experiences partial reflection at a boundary between two media?
7) Name a natural event that produces seismic waves.
8) Which type of seismic wave is longitudinal?
9) Why are S-waves not detected on the opposite side of Earth from which they are generated?

Exam Questions

PRACTICAL

1 A student is investigating the infrared radiation emitted by different surfaces using a Leslie Cube, as shown in **Figure 1**. The student records how long it takes the temperature on each thermometer to increase by 5 °C.

Grade 6-7

Figure 1

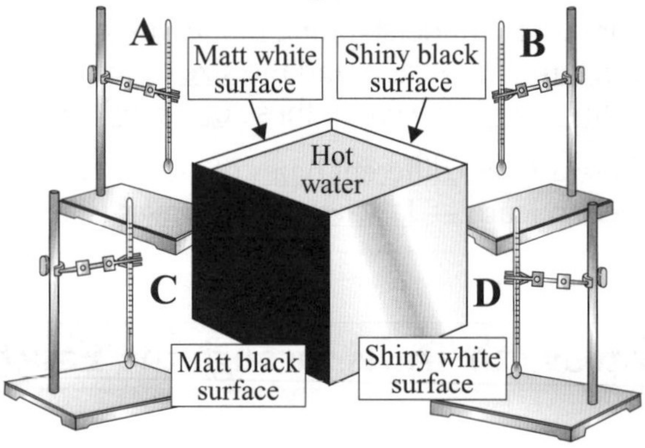

1.1 Suggest **one** thing the student should do to make the experiment a fair test.

[1 mark]

Of the faces of the Leslie Cube, the face at thermometer C behaves most like a perfect black body, and face at thermometer D behaves least like a perfect black body.

1.2 In which of the following orders (fastest to slowest) would you expect the thermometers to increase by 5 °C?
Tick **one** box.

☐ A, B, C, D ☐ C, A, B, D ☐ D, C, B, A ☐ B, A, D, C

[1 mark]

1.3 The water used in the experiment was initially at boiling point, 100 °C.
The experiment is repeated using water at 60 °C.
Predict how this would affect the results. Explain your prediction.

[2 marks]

1.4 Another student suggests that using digital thermometers connected to data loggers to measure the temperatures would improve the investigation.
The digital thermometers measure temperature in °C to two decimal places.
Give **two** reasons why this student is correct.

[2 marks]

Exam Questions

2 Ultrasound waves have frequencies above the normal range of human hearing. *Grade 6-7*

2.1 What is the minimum frequency of an ultrasound wave?

[1 mark]

2.2 An ultrasound transmitter uses mechanical vibrations to produce ultrasound waves.
On **Figure 2**, draw on the direction of the mechanical vibrations.

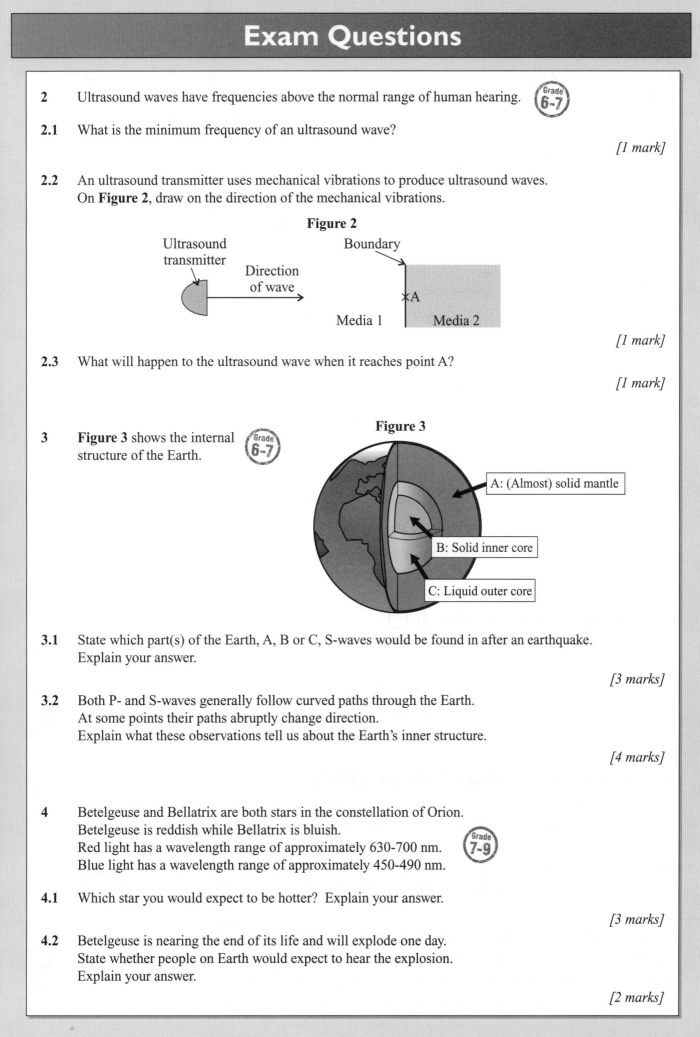

Figure 2

Ultrasound transmitter

Direction of wave

Boundary

×A

Media 1　　Media 2

[1 mark]

2.3 What will happen to the ultrasound wave when it reaches point A?

[1 mark]

3 **Figure 3** shows the internal structure of the Earth. *Grade 6-7*

Figure 3

A: (Almost) solid mantle

B: Solid inner core

C: Liquid outer core

3.1 State which part(s) of the Earth, A, B or C, S-waves would be found in after an earthquake.
Explain your answer.

[3 marks]

3.2 Both P- and S-waves generally follow curved paths through the Earth.
At some points their paths abruptly change direction.
Explain what these observations tell us about the Earth's inner structure.

[4 marks]

4 Betelgeuse and Bellatrix are both stars in the constellation of Orion.
Betelgeuse is reddish while Bellatrix is bluish.
Red light has a wavelength range of approximately 630-700 nm.
Blue light has a wavelength range of approximately 450-490 nm. *Grade 7-9*

4.1 Which star you would expect to be hotter? Explain your answer.

[3 marks]

4.2 Betelgeuse is nearing the end of its life and will explode one day.
State whether people on Earth would expect to hear the explosion.
Explain your answer.

[2 marks]

Revision Summary for Topic 6

And that's the end of Topic 6 — give yourself a pat on the back before seeing how much you've learnt.
- Try these questions and tick off each one when you get it right.
- When you've done all the questions under a heading and are completely happy with it, tick it off.

Wave Properties (p.126-133) ☑

1) What is the amplitude, wavelength, frequency and period of a wave?
2) Describe the difference between transverse and longitudinal waves and give an example of each.
3) Describe an experiment to measure the speed of ripples on water.
4) State a rule that applies to all reflections.
5) Draw a ray diagram for a light ray being reflected where the angle of incidence is 25°.
6) Define specular and diffuse reflection.
7) Draw a diagram showing a light ray crossing, at an angle, into a medium in which it slows down.
8) Describe an experiment you could do to investigate a) refraction and b) reflection of light.

Uses and Dangers of Electromagnetic Waves (p.136-140) ☑

9) True or false? All electromagnetic waves are transverse.
10) What kind of current is used to generate radio waves in an antenna?
11) Explain why microwaves are suitable for satellite communication.
12) Give one use of infrared radiation.
13) What type of radiation is used to transmit a signal in an optical fibre?
14) Name the type of radiation produced by the lamps in tanning beds.
15) What does the term 'ionising radiation' mean?
16) What is 1 Sv in mSv?

Lenses, Colour and Filters (p.142-146) ☑

17) Give the three rules for refraction in a convex lens and the three for a concave lens.
18) Explain the terms 'real image' and 'virtual image'.
19) Draw the ray diagram symbols for a converging lens and a diverging lens.
20) True or false? Opaque objects transmit light.
21) Explain how colour filters can change the colour an object appears.

Infrared Radiation and Black Body Radiation (p.148-151) ☑

22) Compare the IR radiation absorption and emission rates for an object at constant temperature.
23) How could you use a Leslie cube to investigate IR radiation emitted by different surfaces?
24) What happens to radiation hitting a perfect black body?
25) a) Draw a diagram showing radiation reflected, absorbed and emitted by Earth and its atmosphere.
    b) Explain how this absorption, reflection and emission of radiation affects the Earth's temperature.

Sound Waves and Exploring Structures with Waves (p.152-155) ☑

26) What is the frequency range of human hearing?
27) Explain how ultrasound is used in medical imaging, industrial imaging and echo sounding.
28) Describe how S and P waves can be used to explore the structure of the Earth's core.

Magnets

I think magnetism is an <u>attractive</u> subject, but don't get <u>repelled</u> by the exam — <u>revise</u>.

Magnets Produce **Magnetic Fields**

1) All magnets have <u>two poles</u> — <u>north</u> (or north seeking) and <u>south</u> (or south seeking).

2) All magnets produce a <u>magnetic field</u> — a region where <u>other magnets</u> or <u>magnetic materials</u> (e.g. iron, steel, nickel and cobalt) experience a <u>force</u>. (This is a <u>non-contact force</u> — similar to the force on charges in an electric field, like you saw on page 60.)

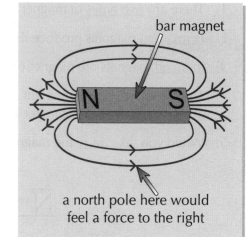

bar magnet

a north pole here would feel a force to the right

3) You can show a magnetic field by drawing <u>magnetic field lines</u>.

4) The lines always go from <u>north to south</u> and they show <u>which way</u> a force would act on a north pole if it was put at that point in the field.

5) The <u>closer together</u> the lines are, the <u>stronger</u> the magnetic field. The <u>further away</u> from a magnet you get, the <u>weaker</u> the field is.

6) The magnetic field is <u>strongest</u> at the <u>poles</u> of a magnet. This means that the <u>magnetic forces</u> are also <u>strongest</u> at the poles.

7) The force between a <u>magnet</u> and a <u>magnetic material</u> is <u>always attractive</u>, no matter the pole.

8) If the two poles of a magnet are put <u>near</u> each other, they will exert a <u>force</u> on each other. This force can be <u>attractive</u> or <u>repulsive</u>. Two poles that are the same (these are called <u>like poles</u>) will <u>repel</u> each other. Two <u>unlike</u> poles will <u>attract</u> each other.

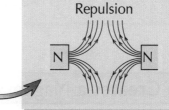

Repulsion   Attraction

Compasses Show the **Direction** of Magnetic Fields

1) Inside a compass is a tiny <u>bar magnet</u> (the needle). The <u>north</u> pole of this magnet is attracted to the south pole of any other magnet it is near. So the compass needle <u>points</u> in the direction of the magnetic field it is in.

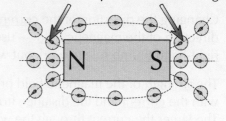

The north pole of the magnet in the compass points along the field line towards the south pole of the bar magnet.

2) You can move a compass around a magnet and <u>trace</u> the needle's position on some paper to build up a picture of what the magnetic field <u>looks like</u>.

3) When they're not near a magnet, compass needles always point <u>north</u>. This is because the <u>Earth</u> generates its own <u>magnetic field</u>, which shows that the <u>inside</u> (<u>core</u>) of the Earth must be <u>magnetic</u>.

A bar magnet's magnetic field lines go from north to south...

...no matter which direction the magnet is pointing. You can see the shape of a <u>magnetic field</u> using <u>compasses</u> or <u>iron filings</u>. Iron filings will give you a pretty pattern — but they won't show you the <u>direction</u> of the <u>magnetic field</u>. That's where compasses really shine. They're also a lot easier to clear up.

Magnetism

Permanent magnets are great, but it would be <u>really</u> handy to be able to turn a magnetic field <u>on</u> and <u>off</u>. Well, it turns out that when <u>electric current</u> flows it <u>produces a magnetic field</u>...

Magnets Can be **Permanent** or **Induced**

1) There are <u>two types</u> of magnet — <u>permanent</u> magnets and <u>induced</u> magnets.

2) <u>Permanent</u> magnets produce their <u>own</u> magnetic field.

3) <u>Induced</u> magnets are magnetic materials that <u>turn into</u> a magnet when they're put into a magnetic field.

4) The force between permanent and induced magnets is always <u>attractive</u> (see magnetic materials on the previous page).

5) When you <u>take away</u> the magnetic field, induced magnets quickly <u>lose</u> most or all of their magnetism.

> N permanent magnet S   magnetic material
>
> The magnetic material becomes magnetised when it is brought near the bar magnet. It has its own poles and magnetic field:
>
> N permanent magnet S   N induced magnet S
>
> induced poles

A **Moving Charge** Creates a Magnetic Field

1) When a <u>current flows</u> through a <u>wire</u>, a <u>magnetic field</u> is created <u>around</u> the wire.

2) The field is made up of <u>concentric circles</u> perpendicular to the wire, with the wire in the centre.

3) You can see this by placing a <u>compass</u> near a <u>wire</u> that is carrying a <u>current</u>. As you move the compass, it will <u>trace</u> the direction of the magnetic field.

4) Changing the <u>direction</u> of the <u>current</u> changes the direction of the <u>magnetic field</u> — use the <u>right-hand thumb rule</u> to work out which way it goes.

5) The <u>strength</u> of the magnetic field produced <u>changes</u> with the <u>current</u> and the <u>distance</u> from the wire. The <u>larger</u> the current through the wire, or the <u>closer</u> to the wire you are, the <u>stronger</u> the field is.

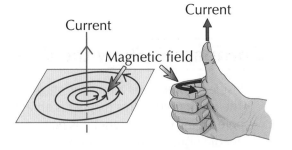

Current

Current

Magnetic field

> <u>The Right-Hand Thumb Rule</u>
> Using your right hand, point your <u>thumb</u> in the direction of <u>current</u> and <u>curl</u> your fingers. The direction of your <u>fingers</u> is the direction of the <u>field</u>.

Just point your thumb in the direction of the current...

...and your fingers show the direction of the field. Remember, it's always your <u>right thumb</u>. Not your left. You'll use your left hand on page 164 though, so it shouldn't feel left out...

Electromagnets

Electric currents can create magnetic fields (see previous page). We can use this to make magnets that can be switched on and off — these are electromagnets.

A **Solenoid** is a Coil of Wire

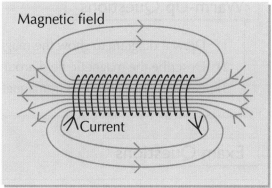

Magnetic field

Current

1) You can increase the strength of the magnetic field that a wire produces by wrapping the wire into a coil called a solenoid.

2) This happens because the field lines around each loop of wire line up with each other. This results in lots of field lines pointing in the same direction that are very close to each other. As you saw on page 159, the closer together field lines are, the stronger the field is.

3) The magnetic field inside a solenoid is strong and uniform (it has the same strength and direction at every point in that region).

4) Outside the coil, the magnetic field is just like the one round a bar magnet.

5) You can increase the field strength of the solenoid even more by putting a block of iron in the centre of the coil. This iron core becomes an induced magnet whenever current is flowing.

6) If you stop the current, the magnetic field disappears.

7) A solenoid with an iron core is called an ELECTROMAGNET (a magnet whose magnetic field can be turned on and off with an electric current).

Electromagnets Have **Lots of Uses**

Magnets you can switch on and off are really useful. They're usually used because they're so quick to turn on and off or because they can create a varying force (like in loudspeakers, p.170).

1) Electromagnets are used in some cranes to attract and pick up things made from magnetic materials like iron and steel, e.g. in scrap yards. Using an electromagnet means the magnet can be switched on when you want to pick stuff up, then switched off when you want to drop it.

2) Electromagnets can also be used within other circuits to act as switches (e.g. in the electric starters of motors), like this:

3) When the switch in circuit one is closed, it turns on the electromagnet, which attracts the iron contact on the rocker.

4) The rocker pivots and closes the contacts, completing circuit two, and turning on the motor.

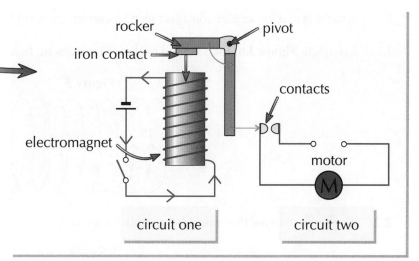

rocker   pivot
iron contact →
contacts
electromagnet
motor
M
circuit one   circuit two

Fields around electromagnets and bar magnets are the same shape

Electromagnets pop up in lots of different places — they're used in electric bells, car ignition circuits and some security doors. Electromagnets aren't all the same strength though — how strong they are depends on stuff like the number of turns of wire there are and the size of the current going through the wire.

Warm-Up & Exam Questions

It's time for another page of questions to check your knowledge retention. If you can do the warm-up questions without breaking into a sweat, then see how you get on with the exam questions below.

Warm-Up Questions

1) Draw a diagram to show the magnetic field around a single bar magnet.
2) Describe the magnetic field around a current-carrying wire.
3) Give one use of an electromagnet.

Exam Questions

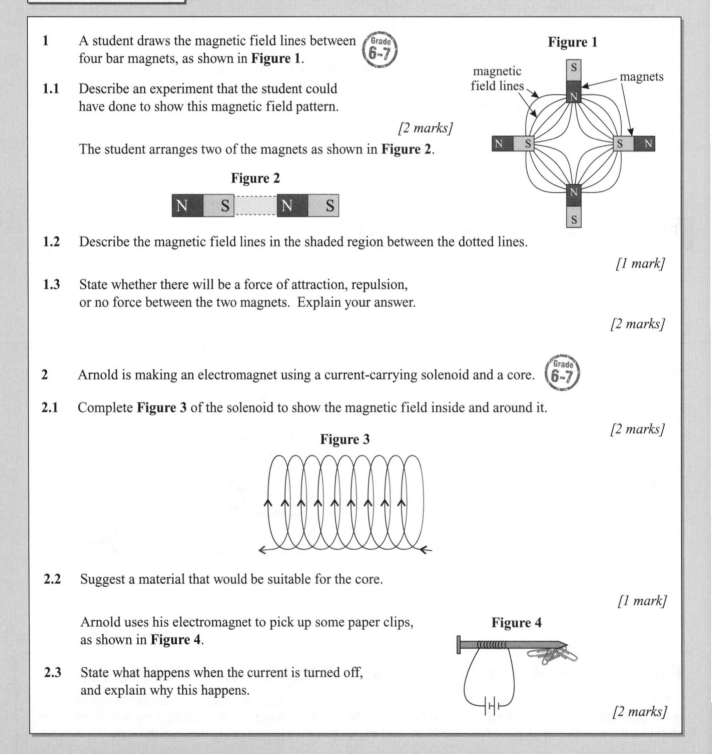

1   A student draws the magnetic field lines between four bar magnets, as shown in **Figure 1**.   Grade 6-7

Figure 1

magnetic field lines ⟶

magnets

1.1   Describe an experiment that the student could have done to show this magnetic field pattern.

[2 marks]

The student arranges two of the magnets as shown in **Figure 2**.

Figure 2

N S   N S

1.2   Describe the magnetic field lines in the shaded region between the dotted lines.

[1 mark]

1.3   State whether there will be a force of attraction, repulsion, or no force between the two magnets. Explain your answer.

[2 marks]

2   Arnold is making an electromagnet using a current-carrying solenoid and a core.   Grade 6-7

2.1   Complete **Figure 3** of the solenoid to show the magnetic field inside and around it.

[2 marks]

Figure 3

2.2   Suggest a material that would be suitable for the core.

[1 mark]

Arnold uses his electromagnet to pick up some paper clips, as shown in **Figure 4**.

Figure 4

2.3   State what happens when the current is turned off, and explain why this happens.

[2 marks]

The Motor Effect

Passing an electric current through a wire produces a magnetic field around the wire (p.160). If you put that wire into a magnetic field, the <u>two magnetic fields interact</u>, which can exert a force on the wire.

A **Current** in a Magnetic Field Experiences a **Force**

When a <u>current-carrying</u> wire (or any other <u>conductor</u>) is put between magnetic poles, the <u>magnetic field</u> around the wire <u>interacts</u> with the magnetic field it has been placed in. This causes the magnet and the conductor to <u>exert a force on each other</u>. This is called the <u>motor effect</u> and can cause the <u>wire</u> to <u>move</u>.

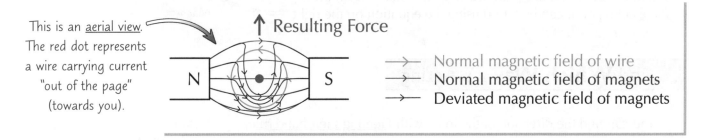

This is an <u>aerial view</u>. The red dot represents a wire carrying current "out of the page" (towards you).

↑ Resulting Force

N   S

→ Normal magnetic field of wire
→ Normal magnetic field of magnets
→ Deviated magnetic field of magnets

1) To experience the <u>full force</u>, the <u>wire</u> has to be at <u>90°</u> to the <u>magnetic field</u>. If the wire runs <u>parallel</u> to the <u>magnetic field</u>, it won't experience <u>any force at all</u>. At angles in between, it'll feel <u>some</u> force.

2) The force always acts at <u>right angles</u> to the <u>magnetic field</u> of the magnets and to the <u>direction of the current</u> in the wire.

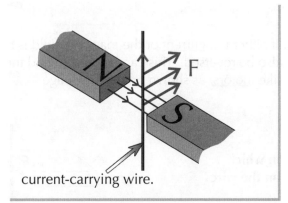

current-carrying wire.

3) A good way of showing the direction of the force is to apply a current to a set of <u>rails</u> inside a <u>horseshoe magnet</u> (shown below). A bar is placed on the rails, which <u>completes the circuit</u>. This generates a <u>force</u> that <u>rolls the bar</u> along the rails.

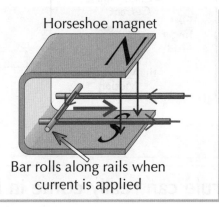

Horseshoe magnet

Bar rolls along rails when current is applied

The motor effect is used in lots of appliances that use movement — see pages 165 and 170.

4) The magnitude (strength) of the force <u>increases</u> with the <u>strength</u> of the <u>magnetic field</u>.

5) The force also <u>increases</u> with the amount of <u>current</u> passing through the conductor.

The Motor Effect

You Can Find the **Size** of the **Force**...

The force acting on a conductor in a magnetic field depends on three things:

1) The magnetic flux density — how many field lines there are in a region.
   This shows the strength of the magnetic field (p.159).

2) The size of the current through the conductor.

3) The length of the conductor that's in the magnetic field.

When the current is at 90° to the magnetic field it is in, the
force acting on it can be found using the equation on the right.

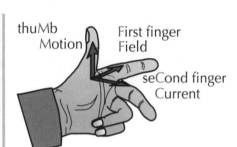

Force (N)   Current (A)
$$F = BIl$$
Length (m)
Magnetic flux density (T, tesla)

... and **Which Way** it's Acting

You can find the direction of the force with Fleming's left-hand rule.

1) Using your left hand, point your First finger
   in the direction of the Field.

2) Point your seCond finger in the direction of the Current.

3) Your thuMb will then point in the
   direction of the force (Motion).

thuMb   First finger
Motion   Field
       seCond finger
       Current

Fleming's left-hand rule shows that if either the current or the magnetic field is reversed,
then the direction of the force will also be reversed. This can be used to find the direction
of the force in all sorts of things — like motors, as shown on the next page.

EXAMPLE:

In the diagram on the right, in which direction does the force act on the wire?

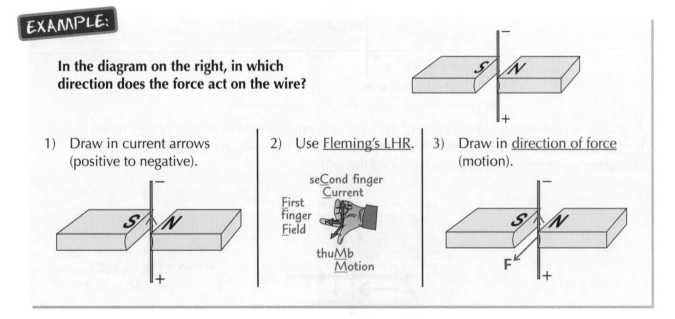

1) Draw in current arrows (positive to negative).

2) Use Fleming's LHR.

seCond finger
Current
First finger Field
thuMb Motion

3) Draw in direction of force (motion).

Fleming's left-hand rule can really come in handy...

EXAM TIP

Use the left-hand rule in the exam. You might look a bit silly, but it makes getting those marks so much easier. So don't get tripped up — your First finger corresponds to the direction of the Field, your seCond finger points in the direction of the Current, and finally your thuMb points in the direction of the force (or Motion). If you can remember that, then those marks will come easily.

Electric Motors

Electric motors use the motor effect (see pages 163-164) to get them (and keep them) moving.
This is one of the favourite exam topics of all time. Read it. Understand it. Learn it. Lecture over.

A Current-Carrying Coil of Wire Rotates in a Magnetic Field

1) The diagram on the right shows a basic dc motor.
    Forces act on the two side arms of a coil of
    wire that's carrying a current.

2) These forces are just the usual forces which
    act on any current in a magnetic field (p.164).

3) Because the coil is on an axle and the
    forces act one up and one down, it rotates.

4) The split-ring commutator is a clever way of
    swapping the contacts every half turn to
    keep the motor rotating in the same direction.

5) The direction of the motor can be reversed either by
    swapping the polarity of the dc supply (reversing the current)
    or swapping the magnetic poles over (reversing the field).

6) You can use Fleming's left-hand rule to work out which way the coil will turn.

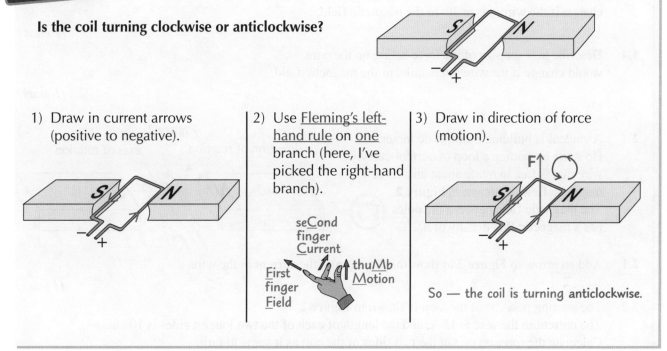

Direct current (dc) is current that only flows in one direction.

EXAMPLE:

Is the coil turning clockwise or anticlockwise?

1) Draw in current arrows
    (positive to negative).

2) Use Fleming's left-
    hand rule on one
    branch (here, I've
    picked the right-hand
    branch).

    seCond
    finger
    Current

    First
    finger
    Field

    thuMb
    Motion

3) Draw in direction of force
    (motion).

    F

    So — the coil is turning anticlockwise.

The motor effect has a lot of important applications...

Electric motors are important components in a lot of everyday items. Food mixers, DVD players, and
anything that has a fan (hair dryers, laptops, etc) use electric motors to keep things turning.

Warm-Up & Exam Questions

Time to test your knowledge — as usual, check you can do the basics, then get stuck into some lovely exam questions. Don't forget to go back and check up on any niggling bits you can't do.

Warm-Up Questions

1) What is the motor effect?
2) In Fleming's left-hand rule, what's represented by the first finger, the second finger and the thumb?
3) Give two changes that can be made to make a dc motor run in reverse.

Exam Questions

1   **Figure 1** shows an aerial view of a current-carrying wire in a magnetic field.
     The circle represents the wire carrying current out of the page, towards you.

Grade 4-6

Figure 1

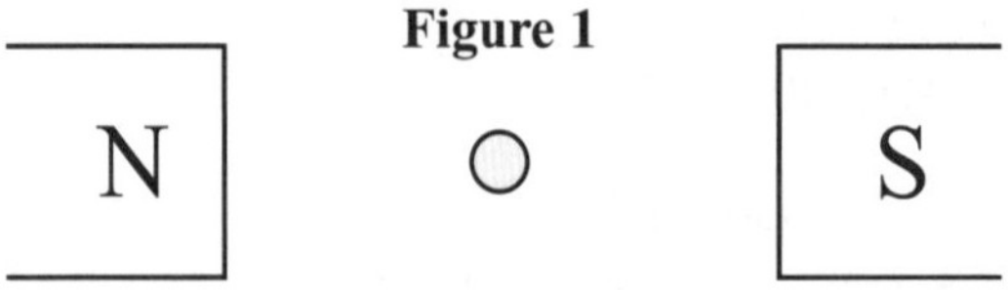

N   ○   S

1.1   On **Figure 1**, draw an arrow to show the direction of the force acting on the current-carrying wire.

[1 mark]

1.2   Describe what would happen to the force acting on the
       current-carrying wire if the direction of the current was reversed.

[1 mark]

1.3   Describe how the size of the force acting on the wire would
       change if the wire was at 30° to the magnetic field.

[1 mark]

1.4   Describe how the size of the force acting on the wire
       would change if the wire ran parallel to the magnetic field.

[1 mark]

2   A student is building a simple dc motor.
     He starts by putting a loop of current-carrying
     wire that is free to rotate about an axis in a
     magnetic field, as shown in **Figure 2**.
     The magnetic field between the poles
     has a magnetic flux density of 0.2 T.

Grade 6-7

Figure 2

direction of rotation   axis of rotation

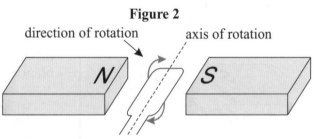

2.1   Add an arrow to **Figure 2** to show the direction of the current in the wire.

[1 mark]

2.2   The starting position of the loop is shown in **Figure 2**.
       The current in the wire is 15 A, and the length of each of the two longest sides is 10 cm.
       Calculate the force on one of the two sides of the coil as it starts to turn.
       Use the correct equation from the Physics Equation Sheet on the inside back cover.

[2 marks]

2.3   The motor will stop rotating in the same direction after 90° of rotation from its start position.
       Suggest and explain how the student could get the motor to keep rotating in the same direction.

[2 marks]

The Generator Effect

Electricity is generated using the <u>generator effect</u> (which is also known as <u>electromagnetic induction</u>). Sounds terrifying, but read this page carefully and hopefully it shouldn't be too complicated.

Cutting Field Lines Induces a Potential Difference

> <u>THE GENERATOR EFFECT</u>: The <u>induction</u> of a <u>potential difference</u> (and <u>current</u> if there's a <u>complete circuit</u>) in a wire which is experiencing a <u>change in magnetic field</u>.

1) The <u>generator effect</u> induces (creates) a <u>potential difference</u> in a conductor (and a <u>current</u> if the conductor is part of a <u>complete circuit</u>).

2) You can do this by moving a <u>magnet</u> in a <u>coil of wire</u>...

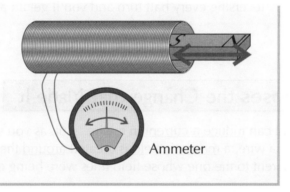

Ammeter

...OR moving a <u>conductor</u> (wire) in a <u>magnetic field</u> ("cutting" magnetic field lines).

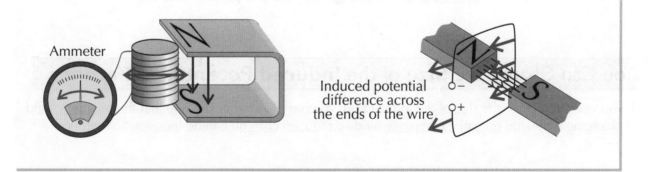

Ammeter

Induced potential difference across the ends of the wire

3) Shifting the magnet from <u>side to side</u> creates a little "<u>blip</u>" of current in the conductor if it's part of a <u>complete circuit</u> (the current can be shown on an ammeter in the circuit).

4) If you move the magnet (or conductor) in the <u>opposite direction</u>, then the potential difference/current will be <u>reversed</u>. Likewise if the <u>polarity</u> of the magnet is <u>reversed</u>, then the potential difference/current will be <u>reversed</u> too.

5) If you keep the magnet (or the coil) moving <u>backwards and forwards</u>, you produce a potential difference that keeps swapping direction, which produces an <u>alternating current</u>.

The Generator Effect

Time for more on the generator effect and how <u>rotation</u> can cause the <u>generator effect</u>. Fascinating stuff.

Rotation Can Also Cause the **Generator Effect**

You can create the same effect by turning a magnet <u>end to end</u> in a coil, or turning a coil inside a magnetic field. This is how <u>generators</u> work to produce <u>alternating current</u> (<u>ac</u>) or <u>direct current</u> (<u>dc</u>) — see next page.

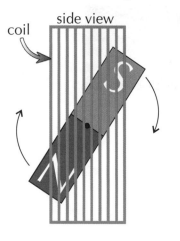

side view

coil

1) As you turn the magnet, the magnetic field through the coil <u>changes</u>. This change in the magnetic field induces a <u>potential difference</u>, which can make a <u>current</u> flow in the wire.

2) Every time the magnet moves through half a turn, the <u>direction</u> of the magnetic field through the coil <u>reverses</u>. When this happens, the potential difference reverses, so the <u>current</u> flows in the <u>opposite direction</u> around the coil of wire.

3) If you keep turning the magnet in the <u>same direction</u> — always clockwise, say — then the potential difference will keep on reversing every half turn and you'll get an <u>alternating current</u>.

Induced Current **Opposes** the Change that Made It

1) So, a change in magnetic field can <u>induce a current</u> in a wire. But, as you saw on page 160, when a current flows through a wire, a <u>magnetic field</u> is created <u>around</u> the wire. (Yep, that's a <u>second</u> magnetic field — different to the one whose field lines were being cut in the first place.)

2) The <u>magnetic field</u> created by an <u>induced</u> current always acts <u>against the change</u> that made it (whether that's the movement of a wire or a change in the field it's in). Basically, it's trying to return things to <u>the way they were</u>.

3) This means that the <u>induced current</u> always <u>opposes</u> the change that made it.

You Can Change the **Size** of the **Induced Potential Difference**

If you want to change the <u>size</u> of the induced pd, you have to change the <u>rate</u> that the <u>magnetic field</u> is changing. Induced <u>potential difference</u> (and so <u>induced current</u>) can be <u>increased</u> by either:

1) Increasing the <u>speed</u> of the movement — cutting <u>more</u> magnetic field lines in a given <u>time</u>.

2) Increasing the <u>strength</u> of the <u>magnetic field</u> (so there are more field lines that can be cut).

The generator effect — works whether the coil or the field is moving

It doesn't matter what's moving, the <u>generator effect</u> will work as long as field lines are being 'cut'. Remember, the current that's induced will 'oppose' the change that generated it.

Alternators and Dynamos

Generators make use of the generator effect to induce a current. Whether this current is alternating or direct depends on how the coil is connected to the circuit. Don't get the types of connection mixed up.

Alternators Generate Alternating Current

1) Generators rotate a coil in a magnetic field (or a magnet in a coil).

2) Their construction is pretty much like a motor.

3) As the coil (or magnet) spins, a current is induced in the coil. This current changes direction every half turn.

4) Instead of a split-ring commutator, alternators have slip rings and brushes so the contacts don't swap every half turn.

5) This means they produce an alternating potential difference (pd) — more on this below.

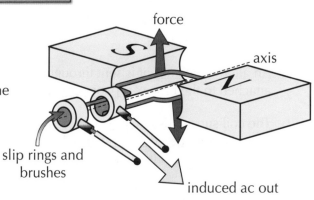

force
axis
slip rings and brushes
induced ac out

Dynamos Generate Direct Current

1) Dynamos work in the same way as alternators, apart from one important difference.

2) They have a split-ring commutator instead of slip rings.

3) This swaps the connection every half turn to keep the current flowing in the same direction (similar to in a dc motor, p.165).

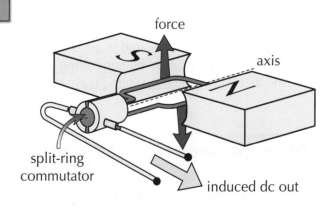

force
axis
split-ring commutator
induced dc out

You Can Use an Oscilloscope To See the Generated pd

1) Oscilloscopes show how the potential difference generated in the coil changes over time.

2) For ac this is a line that goes up and down, crossing the horizontal axis.

3) For dc the line isn't straight like you might expect, but it stays above the axis (the pd is always positive) so it's still direct current.

4) The height of the line at a given point is the generated potential difference at that time.

5) Increasing the frequency of revolutions increases the overall pd, but it also creates more peaks too.

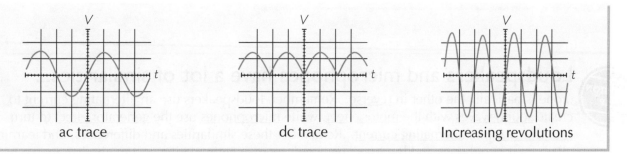

ac trace   dc trace   Increasing revolutions

Loudspeakers and Microphones

If you've ever broken a pair of headphones, you'll have seen the tiny crinkly paper cone inside them. I'm sure you've never sat and wondered how they work, but it's all down to electromagnetism...

Loudspeakers Work Because of the Motor Effect

Loudspeakers and headphones (which are just tiny loudspeakers) both use electromagnets:

1) An alternating current is sent through a coil of wire attached to the base of a paper cone.

2) The coil surrounds one pole of a permanent magnet, and is surrounded by the other pole, so the current causes a force on the coil (which causes the cone to move).

3) When the current reverses, the force acts in the opposite direction, which causes the cone to move in the opposite direction too.

4) So variations in the current make the cone vibrate, which makes the air around the cone vibrate and creates the variations in pressure that cause a sound wave (p.152).

5) The frequency of the sound wave is the same as the frequency of the ac, so by controlling the frequency of the ac you can alter the sound wave produced.

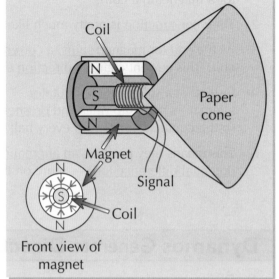

Front view of magnet

Microphones Generate Current From Sound Waves

1) Microphones are basically loudspeakers in reverse.

2) Sound waves hit a flexible diaphragm that is attached to a coil of wire, wrapped around a magnet.

3) This causes the coil of wire to move in the magnetic field, which generates a current.

4) The movement of the coil (and so the generated current) depends on the properties of the sound wave (louder sounds make the diaphragm move further).

5) This is how microphones can convert the pressure variations of a sound wave into variations in current in an electric circuit.

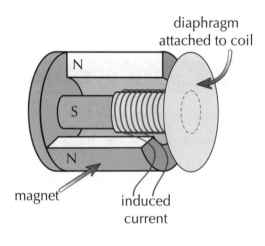

diaphragm attached to coil

magnet

induced current

Loudspeakers and microphones have a lot of similarities...

...one is basically the other in reverse. Remember, loudspeakers use an alternating current to create sound waves with the motor effect, while microphones use the generator effect to turn sound waves into alternating current. Remember these similarities and differences, and learning all about microphones and loudspeakers won't seem as intimidating as it might first appear.

Transformers

Transformers use the generator effect to change the size of a potential difference (pd).
They will only work with alternating current.

Transformers Change the pd — but Only for **Alternating** Current

1) Transformers change the size of the potential difference of an alternating current.

2) They all have two coils of wire, the primary and the secondary, joined with an iron core.

3) When an alternating pd is applied across the primary coil, the iron core magnetises and demagnetises quickly. This changing magnetic field induces an alternating pd in the secondary coil.

4) If the second coil is part of a complete circuit, this causes a current to be induced.

5) The ratio between the primary and secondary potential differences is the same as the ratio between the number of turns on the primary and secondary coils.

Iron is used because it's easily magnetised.

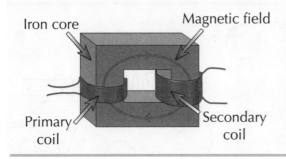

Iron core   Magnetic field

Primary coil   Secondary coil

STEP-UP TRANSFORMERS step the potential difference up (i.e. increase it). They have more turns on the secondary coil than the primary coil.

STEP-DOWN TRANSFORMERS step the potential difference down (i.e. decrease it). They have more turns on the primary coil than the secondary.

The **Transformer Equation** — Use it Either Way Up

1) As long as you know the input pd and the number of turns on each coil, you can calculate the output pd from a transformer using the transformer equation:

V_p = input potential difference
V_s = output potential difference

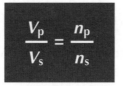
$$\frac{V_p}{V_s} = \frac{n_p}{n_s}$$

n_p = number of turns on primary coil
n_s = number of turns on secondary coil

So for a step-up transformer, $V_s > V_p$ and for a step-down transformer, $V_s < V_p$.

2) This equation can be used either way up, so  $\frac{V_s}{V_p} = \frac{n_s}{n_p}$ works just as well.

There's less rearranging to do if you put whatever you're trying to find (the unknown) on the top.

3) Transformers are almost 100% efficient. So you can assume that the input power is equal to the output power. Using $P = VI$ from page 54, you can write the equation you saw on page 56:

V_s = pd across secondary coil
I_s = current through secondary coil

$$V_s I_s = V_p I_p$$

V_p = pd across primary coil
I_p = current through primary coil

Remember, $V_s \times I_s$ is the power output at the secondary coil and $V_p \times I_p$ is the power input at the primary coil.

4) You need to be able to relate both of these equations to power transmission in the national grid, to explain why and how the national grid transmits at very high pds.

5) You've already seen on page 56 that a low current means that less energy is wasted heating the wires and the surroundings, making the national grid an efficient way of transmitting power. The second equation shows why, for a given power, a high pd is needed for a low current.

6) The first equation can be used to work out the number of turns needed to increase the pd (and decrease the current) to the right levels.

Warm-Up & Exam Questions

There were lots of new ideas in that section, not to mention those equations on page 171. Better have a go at these questions so you can really see what's gone in and what you might need to go over again.

Warm-Up Questions

1) State two ways in which you can increase the pd induced by the generator effect.
2) What type of current does a dynamo generate?
3) Do step-up transformers have more turns on their primary or secondary coil?

Exam Questions

1 **Figure 1** shows a coil of wire connected to an ammeter. Ravi moves a bar magnet into the coil as shown. The pointer on the ammeter moves to the left.

Grade 4-6

Figure 1

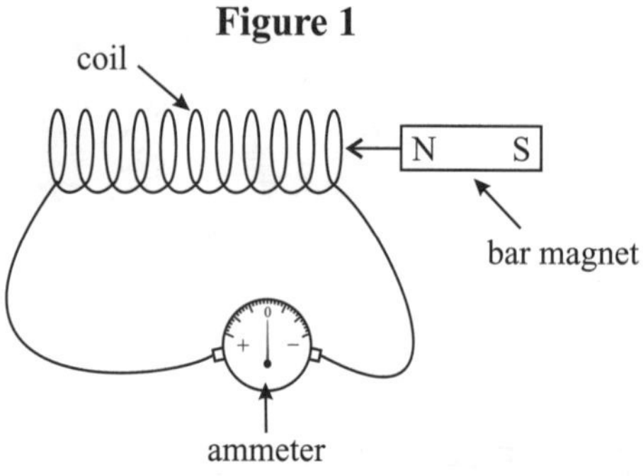

1.1 Explain why the pointer moves.

[1 mark]

1.2 Suggest how Ravi could get the ammeter's pointer to move to the right.

[1 mark]

1.3 Suggest how Ravi could get a larger reading on the ammeter.

[1 mark]

1.4 What reading will the ammeter show if Ravi holds the magnet still inside the coil?

[1 mark]

2 A student is trying to test a transformer using a dc power supply.

Grade 6-7

2.1 Explain why the voltmeter connected to the secondary coil reads 0 V.

[3 marks]

The student finds an ac power supply and reconnects the transformer.
She finds that: $V_P = 12$ V, $I_P = 2.5$ A, $V_S = 4$ V (where V_P is the pd across the primary coil, etc.).

2.2 Calculate the power input to the transformer.

[2 marks]

2.3 Calculate the current in the secondary coil, I_S.

[4 marks]

2.4 The primary coil has 15 turns. How many turns must be on the secondary coil? Use the correct equation from the Physics Equation Sheet on the inside back cover.

[3 marks]

Figure 2

3 **Figure 2** shows the parts inside an earphone. Sound waves are caused by mechanical vibrations. Explain how the earphone uses an ac supply to produce sound waves.

Grade 7-9

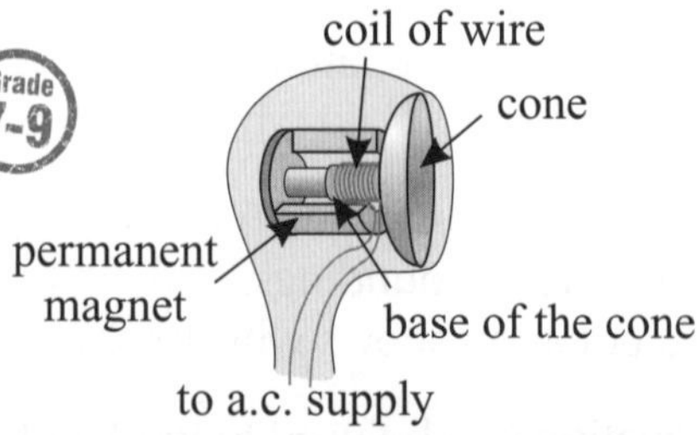

[4 marks]

Revision Summary for Topic 7

That wraps up <u>Topic 7</u> — time to put yourself to the test and find out <u>how much you really know</u>.
- Try these questions and <u>tick off each one</u> when you <u>get it right</u>.
- When you've done <u>all the questions</u> under a heading and are <u>completely happy</u> with it, tick it off.

Magnetism and Basic Electromagnetism (p.159-161) ☐

1) What is a magnetic field?
2) Give three magnetic materials.
3) In what direction do magnetic field lines point?
4) True or false? The force between a magnet and a magnetic material is always repulsive.
5) Describe how you could use a compass to show the direction of a bar magnet's magnetic field lines.
6) Describe the behaviour of a compass needle that is far away from any magnets.
7) What happens to an induced magnet when it is moved far away from a permanent magnet?
8) How can you work out the direction of the magnetic field around a current-carrying wire without using a compass?
9) Why does adding an iron core to a solenoid increase the strength of its magnetic field?
10) Describe an electromagnet.

The Motor Effect (p.163-165) ☑

11) Explain why a current-carrying conductor in a magnetic field experiences a force.
12) State the equation for calculating the size of this force.
13) Name three ways you could increase the force on a current-carrying wire in a magnetic field.
14) What is Fleming's left-hand rule used for?
15) Explain how a basic dc motor works.

The Generator Effect, Transformers and Other Applications (p.167-171) ☐

16) Describe how you can induce a current in a wire, using a magnetic field.
17) Give two ways you could reverse the direction of an induced current.
18) True or false? Induced currents create magnetic fields that oppose the change that made them.
19) Which type of generator uses slip rings and brushes?
20) Draw a graph of potential difference against time for an ac supply.
21) Draw the magnetic field lines for the magnet inside a loudspeaker.
22) Explain how microphones translate sound waves into electrical signals.
23) What kind of current do transformers use?
24) Why do transformers have a core of iron?
25) True or false?
    Step-down transformers have more coils on their primary coil than on their secondary coil.
26) A transformer has an input pd of 100 V and an output pd of 20 V.
    What kind of transformer is it?
27) Write down the equation that relates the input and output currents and pds of transformers.
    What assumption is made when using this equation?
28) Explain how transformers are used to improve efficiency when transmitting electricity.

The Life Cycle of Stars

Stars go through <u>many traumatic stages</u> in their lives — just like teenagers.

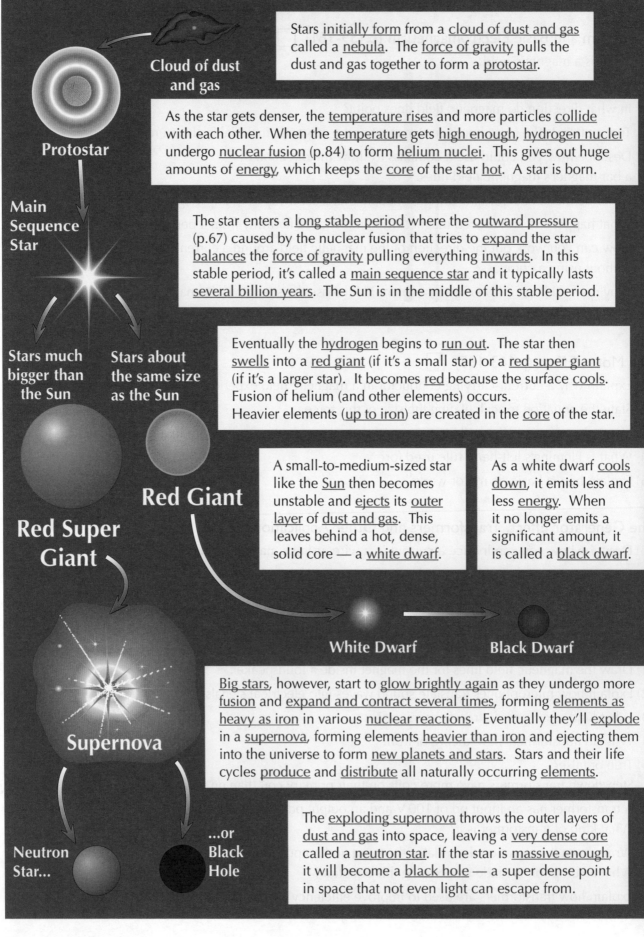

Cloud of dust and gas

Stars <u>initially form</u> from a <u>cloud of dust and gas</u> called a <u>nebula</u>. The <u>force of gravity</u> pulls the dust and gas together to form a <u>protostar</u>.

Protostar

As the star gets denser, the <u>temperature rises</u> and more particles <u>collide</u> with each other. When the <u>temperature</u> gets <u>high enough</u>, <u>hydrogen nuclei</u> undergo <u>nuclear fusion</u> (p.84) to form <u>helium nuclei</u>. This gives out huge amounts of <u>energy</u>, which keeps the <u>core</u> of the star <u>hot</u>. A star is born.

Main Sequence Star

The star enters a <u>long stable period</u> where the <u>outward pressure</u> (p.67) caused by the nuclear fusion that tries to <u>expand</u> the star <u>balances</u> the <u>force of gravity</u> pulling everything <u>inwards</u>. In this stable period, it's called a <u>main sequence star</u> and it typically lasts <u>several billion years</u>. The Sun is in the middle of this stable period.

Stars much bigger than the Sun   **Stars about the same size as the Sun**

Eventually the <u>hydrogen</u> begins to <u>run out</u>. The star then <u>swells</u> into a <u>red giant</u> (if it's a small star) or a <u>red super giant</u> (if it's a larger star). It becomes <u>red</u> because the surface <u>cools</u>. Fusion of helium (and other elements) occurs. Heavier elements (<u>up to iron</u>) are created in the <u>core</u> of the star.

Red Giant

Red Super Giant

A small-to-medium-sized star like the <u>Sun</u> then becomes unstable and <u>ejects</u> its <u>outer layer</u> of <u>dust and gas</u>. This leaves behind a hot, dense, solid core — a <u>white dwarf</u>.

As a white dwarf <u>cools down</u>, it emits less and less <u>energy</u>. When it no longer emits a significant amount, it is called a <u>black dwarf</u>.

White Dwarf   **Black Dwarf**

<u>Big stars</u>, however, start to <u>glow brightly again</u> as they undergo more <u>fusion</u> and <u>expand and contract several times</u>, forming <u>elements as heavy as iron</u> in various <u>nuclear reactions</u>. Eventually they'll <u>explode</u> in a <u>supernova</u>, forming elements <u>heavier than iron</u> and ejecting them into the universe to form <u>new planets and stars</u>. Stars and their life cycles <u>produce</u> and <u>distribute</u> all naturally occurring <u>elements</u>.

Supernova

Neutron Star...   **...or Black Hole**

The <u>exploding supernova</u> throws the outer layers of <u>dust and gas</u> into space, leaving a <u>very dense core</u> called a <u>neutron star</u>. If the star is <u>massive enough</u>, it will become a <u>black hole</u> — a super dense point in space that not even light can escape from.

The Solar System

The <u>Sun</u> is the centre of our <u>solar system</u>. It's <u>orbited</u> by <u>eight planets</u>, along with a bunch of other objects.

Our Solar System has **One** Star — The **Sun**

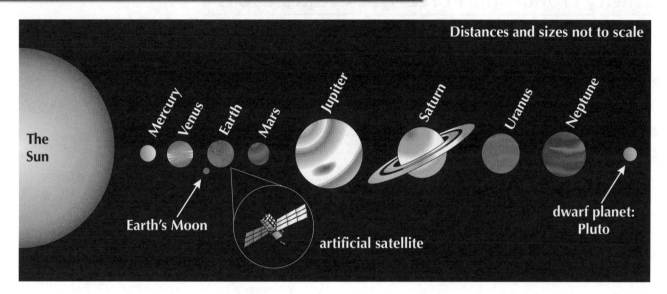

Distances and sizes not to scale

The Sun
Mercury
Venus
Earth
Mars
Jupiter
Saturn
Uranus
Neptune

Earth's Moon

artificial satellite

dwarf planet: Pluto

The <u>solar system</u> is <u>the Sun</u> and all the <u>stuff</u> that <u>orbits around it</u>. This includes things like:

1) <u>Planets</u> — these are large objects that <u>orbit a star</u>. There are <u>eight</u> in our solar system. They also have to be large enough to have "<u>cleared their neighbourhoods</u>". This means that their gravity is strong enough to have <u>pulled in</u> any nearby objects apart from their <u>satellites</u>.

2) <u>Dwarf planets</u>, like our pal Pluto. These are planet-like objects that orbit stars, but don't meet all of the rules for being a planet.

3) <u>Moons</u> — these orbit <u>planets</u>. They're a type of <u>natural satellite</u> (i.e. they're not man-made).

   A satellite is an object that orbits a second, more massive object.

4) <u>Artificial satellites</u> are satellites that humans have built. They generally orbit the <u>Earth</u>.

You are here

Our solar system is a tiny part of the <u>Milky Way galaxy</u>. This is a <u>massive</u> collection of <u>billions</u> of stars that are all held together by gravity.

The solar system — the Sun and everything that moves around it...

Make sure you know the <u>key differences</u> between each type of object in the solar system really well — particularly what each type of object <u>orbits</u>. Speaking of orbits, the next page is all about them.

Orbits

The structure of the <u>solar system</u> is determined by <u>orbits</u> — the paths that objects take as they move around each other in space. I bet you can't wait to find out more. Well, read on...

Gravity Provides the Force That Creates Orbits

1) The planets move around the Sun in <u>almost circular</u> orbits (the same goes for the <u>Moon</u> orbiting the <u>Earth</u>).

2) If an object is <u>travelling in a circle</u> it is <u>constantly changing direction</u>, which means it is <u>constantly accelerating</u>. (Just like a car going around a roundabout.)

3) This also means it is <u>constantly changing velocity</u> (but <u>NOT</u> changing <u>speed</u>).

4) For an object to accelerate, there <u>must</u> be a <u>force</u> acting on it (p.111). This force is directed towards the <u>centre</u> of the circle.

5) This force would cause the object to just <u>fall</u> towards whatever it was orbiting, but as the object is <u>already moving</u>, it just causes it to <u>change its direction</u>.

6) The object <u>keeps accelerating</u> towards what it's orbiting but the <u>instantaneous velocity</u> (which is at a <u>right angle</u> to the acceleration) keeps it travelling in a <u>circle</u>.

7) The force that makes this happen is provided by the <u>gravitational force</u> (gravity) between the <u>planet</u> and the Sun (or between the <u>planet</u> and its <u>satellites</u>).

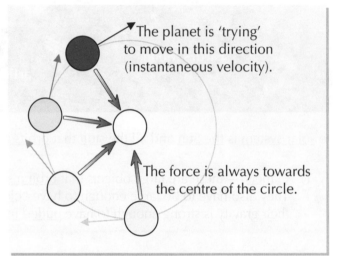

The planet is 'trying' to move in this direction (instantaneous velocity).

The force is always towards the centre of the circle.

The Size of the Orbit Depends on the Object's Speed

1) The <u>closer</u> you get to a star or planet, the <u>stronger</u> the <u>gravitational force</u> is.

2) The <u>stronger</u> the force, the <u>faster</u> the orbiting object needs to travel to remain in <u>orbit</u> (to not crash into the object that it's orbiting).

3) For an object in a <u>stable orbit</u>, if the <u>speed</u> of the object <u>changes</u>, the <u>size</u> (radius) of its <u>orbit</u> must do so too. <u>Faster</u> moving objects will move in a <u>stable</u> orbit with a <u>smaller radius</u> than <u>slower</u> moving ones.

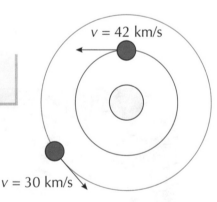

v = 42 km/s

v = 30 km/s

Orbits — when things keep going round in (almost) circles...

Objects are kept in orbit by <u>gravitational attraction</u> causing an acceleration towards the centre of the orbit. Remember, <u>acceleration</u> always <u>changes an object's velocity</u>, but it doesn't necessarily change its <u>speed</u>. (It's all about the difference between vector and scalar quantities. Check out page 87 for more on this.) Now turn over for the final page in this topic. Not far to go now...

Red-shift and the Big Bang

'How it all began' is a tricky question that we just can't answer. Our <u>best guess</u> at the minute is the <u>Big Bang</u>.

The Universe Seems to Be **Expanding**

As big as the universe already is, it looks like it's getting even <u>bigger</u>.
All its <u>galaxies</u> seem to be <u>moving away</u> from each other. There's good evidence for this...

1) When we look at <u>light from most distant galaxies</u>, we find that the <u>wavelength</u> has increased.

2) The wavelengths are all <u>longer</u> than they should be — they're <u>shifted</u> towards the <u>red end</u> of the spectrum. This is called <u>red-shift</u>.

3) This suggests the <u>source</u> of the light is <u>moving away</u> from us. <u>Measurements</u> of the red-shift indicate that these <u>distant galaxies</u> are <u>moving away from us</u> (receding) very quickly — and it's the <u>same result</u> whichever direction you look in.

4) <u>More distant</u> galaxies have <u>greater</u> red-shifts than nearer ones. This means that more distant galaxies are <u>moving away faster</u> than nearer ones — and so <u>all galaxies</u> are <u>moving away</u> from <u>every other galaxy</u>, not just ours.

   You can think of the light wave 'stretching out' as the source moves away from us.

5) The inescapable <u>conclusion</u> appears to be that the whole universe (space itself) is <u>expanding</u>.

> Imagine a <u>balloon</u> covered with <u>pompoms</u>.
> As you <u>blow</u> into the balloon, it <u>stretches</u>.
> The pompoms move <u>further away</u> from each other.
>
> The balloon represents the <u>universe</u> and each pompom is a <u>galaxy</u>. As time goes on, <u>space stretches</u> and expands, moving the galaxies away from each other.
>
> This is a <u>simple model</u> (balloons only stretch <u>so far</u>, and there would be galaxies '<u>inside</u>' the balloon too) but it shows how the <u>expansion</u> of space makes it look like galaxies are <u>moving away</u> from us.

This Evidence Suggests the Universe **Started** with a **Bang**

So all the galaxies are moving away from each other at great speed — suggesting something must have <u>got them going</u>. That 'something' was probably a <u>big explosion</u> — the <u>Big Bang</u>. Here's the theory...

1) Initially, all the matter in the universe occupied <u>a very small space</u>. This tiny space was very <u>dense</u> and so was very <u>hot</u>.

2) Then it '<u>exploded</u>' — space started expanding, and the <u>expansion</u> is still going on.

New **Evidence** Might **Change** Our Theories

1) Something important to remember is that the Big Bang theory is the best guess we have <u>so far</u>. Whenever scientists discover <u>new evidence</u>, they have to either make a <u>new theory</u> or <u>change</u> a current one to <u>explain</u> what they've observed.

2) There is still <u>lots</u> we don't know about the universe. <u>Observations of supernovae</u> from 1998 to the present day appear to show that <u>distant galaxies</u> are moving away from us <u>faster</u> and <u>faster</u> (the <u>speed</u> at which they're receding is <u>increasing</u>).

3) Currently scientists think the universe is mostly made up of <u>dark matter</u> and <u>dark energy</u>. Dark matter is the name given to an <u>unknown substance</u> which holds galaxies <u>together</u>, but does not emit any <u>electromagnetic radiation</u>. Dark energy is thought to be responsible for the <u>accelerated expansion</u> of the universe. But no-one really knows <u>what these things are</u>, so there are lots of different <u>theories</u> about it. These theories get <u>tested</u> over time and are either accepted or rejected.

Warm-Up & Exam Questions

You know the drill — some warm-up questions to get you thinking about what you've just read, and some exam questions to see how well you can apply your knowledge. Get cracking.

Warm-Up Questions

1) What are protostars formed from?
2) Immediately after the end of its main sequence phase, what does a star much larger than the Sun become?
3) Name two types of object in the solar system that orbit the Sun.
4) What is the force responsible for keeping planets in orbit?
5) What is the name given to the observed increase in the wavelength of light from distant stars?
6) State two possible components of the universe which cannot currently be explained.

Exam Questions

1   Use the correct words from the box below to complete the following sentences.   (Grade 4-6)

artificial satellites	planets	the Sun	Jupiter
the Earth	natural satellites	stars	

Moons are .... They orbit ....

.... are man-made, and most of them orbit ....

[4 marks]

2   Stars go through many stages in their lives.   (Grade 6-7)

2.1   Describe how a star is formed.

[3 marks]

2.2   The stable period of a main sequence star can last billions of years.
Explain why main sequence stars undergo a stable period.

[2 marks]

When main sequence stars begin to run out of hydrogen in their core, they swell and become either a red giant or a red super giant depending on their size.

2.3   Describe what happens to stars after their red giant phase until the end of their life cycle.

[3 marks]

2.4   Describe what happens to stars after their red super giant phase until the end of their life cycle.

[3 marks]

3*   Describe red-shift and explain how it supports the Big Bang   (Grade 7-9)
theory as an explanation for how the universe began.

[6 marks]

Revision Summary for Topic 8

<u>Topic 8</u> — short, sweet and super interesting. Now it's time to check you filled all that space in your head.
- Try these questions and <u>tick off each one</u> when you <u>get it right</u>.
- When you've done <u>all the questions</u> under a heading and are <u>completely happy</u> with it, tick it off.

The Life Cycle of Stars (p.174) ☑
1) What is a nebula?
2) What causes the rise in temperature that leads to nuclear fusion in a protostar?
3) What causes a main sequence star to remain stable for a long time?.
4) Which part of its life cycle is our Sun currently in?
5) What happens to a star about the same size as our Sun when it begins to run out of hydrogen?
6) What is a black dwarf and how is it made?
7) At what stage in a star's life cycle are elements heavier than iron formed?
8) True or false? The Sun will eventually turn into a black hole.

The Solar System (p.175) ☑
9) How many stars are there in our solar system?
10) What do planets and dwarf planets orbit?
11) True or false? Pluto is a dwarf planet.
12) What galaxy is our solar system part of?

Orbits (p.176) ☑
13) What is the approximate shape of the planets' orbits around the Sun?
14) True or false? An object in a stable orbit has a continually changing speed.
15) What is the name of the force that pulls an orbiting object towards Earth?
16) State the direction of the orbiting object's instantaneous velocity in relation to this force.
17) How does the strength of a planet's gravitational force change as you get closer to its surface?
18) What is the relationship between the speed of an orbiting object and its orbital radius?

Red-shift and the Big Bang (p.177) ☑
19) What is red-shift?
20) True or false? Very distant galaxies are moving away faster than ones closer to us.
21) Give two limitations of the balloon model of the expanding universe.
22) Briefly describe the Big Bang theory.
23) What did scientists discover about the movement of galaxies in 1998 and how did they discover this?
24) True or false? New evidence that disproves a popular theory is ignored.

Measuring Lengths and Angles

Get your lab coat on, it's time to find out about the skills you'll need in <u>experiments</u>.
First things first — make sure you're using <u>appropriate equipment</u> and know <u>how to use it</u> correctly.

Measure **Most Lengths** with a **Ruler**

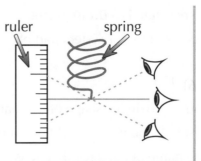

1) In most cases a bog-standard <u>centimetre ruler</u> can be used to measure <u>length</u>. It depends on what you're measuring though — <u>metre rulers</u> are handy for <u>large</u> distances, while <u>micrometers</u> are used for measuring tiny things like the <u>diameter of a wire</u>.

2) The ruler should always be <u>parallel to</u> what you want to measure.

3) If you're dealing with something where it's <u>tricky</u> to measure just <u>one</u> accurately (e.g. water ripples, p.128), you can measure the length of <u>some</u> of them and then <u>divide</u> to find the <u>length of one</u>.

4) If you're taking <u>multiple measurements</u> of the <u>same</u> object (e.g. to measure changes in length) then make sure you always measure from the <u>same point</u> on the object. It can help to draw or stick small <u>markers</u> onto the object to line up your ruler against.

5) Make sure the ruler and the object are always at <u>eye level</u> when you take a reading. This stops <u>parallax</u> affecting your results.

Parallax is where a measurement appears to <u>change</u> based on <u>where you're looking from</u>.

The <u>blue line</u> is the measurement taken when the spring is at <u>eye level</u>. It shows the correct length of the spring.

Use a **Protractor** to Find **Angles**

1) First align the <u>vertex</u> (point) of the angle with the mark in the <u>centre</u> of the protractor.

2) Line up the <u>base line</u> of the protractor with one line that forms the <u>angle</u> and then measure the angle of the other line using the scale on the <u>protractor</u>.

3) If the lines creating the angle are very <u>thick</u>, align the protractor and measure the angle from the <u>centre</u> of the lines. Using a <u>sharp pencil</u> to draw diagrams helps to <u>reduce errors</u> when measuring angles.

4) If the lines are <u>too short</u> to measure easily, you may have to <u>extend</u> them. Again, make sure you use a <u>sharp pencil</u> to do this.

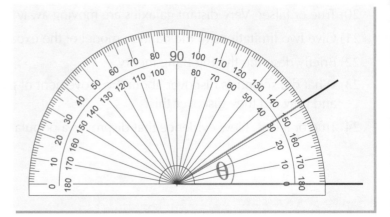

Measuring Volumes

Did you order some more measuring? Well, even if you didn't, here's some stuff about <u>measuring volumes</u>.

Measuring Cylinders and Pipettes Measure Liquid Volumes

1) <u>Measuring cylinders</u> are the most common way to measure a liquid.

2) They come in all different <u>sizes</u>. Make sure you choose one that's the <u>right size</u> for the measurement you want to make. It's no good using a huge 1 dm³ cylinder to measure out 2 cm³ of a liquid — the graduations (markings for scale) will be <u>too big</u> and you'll end up with <u>massive errors</u>. It'd be much better to use one that measures up to 10 cm³.

3) You can also use a <u>pipette</u> to measure volume. <u>Pipettes</u> are used to suck up and <u>transfer</u> volumes of liquid between containers.

4) <u>Graduated pipettes</u> are used to transfer <u>accurate</u> volumes. A <u>pipette filler</u> is attached to the end of a graduated pipette to <u>control</u> the amount of liquid being drawn up.

5) Whichever method you use, always read the volume from the <u>bottom of the meniscus</u> (the curved upper surface of the liquid) when it's at <u>eye level</u>.

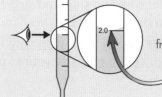

Read volume from here — the bottom of the meniscus.

Eureka Cans Measure the Volumes of Solids

1) <u>Eureka cans</u> are used in <u>combination</u> with <u>measuring cylinders</u> to find the volumes of <u>irregular solids</u> (p.64)

2) They're essentially a <u>beaker with a spout</u>. To use them, fill them with water so the water level is <u>above the spout</u>.

3) Let the water <u>drain</u> from the spout, leaving the water level <u>just below</u> the start of the spout (so <u>all</u> the water displaced by an object goes into the measuring cylinder and gives you the <u>correct volume</u>).

4) Place a <u>measuring cylinder</u> below the end of the spout. When you place a solid in the beaker, it causes the water level to <u>rise</u> and water to flow out of the spout.

5) Make sure you wait until the spout has <u>stopped dripping</u> before you measure the volume of the water in the measuring cylinder. The object's <u>volume</u> is equal to the <u>volume of the water</u> in the <u>measuring cylinder</u>.

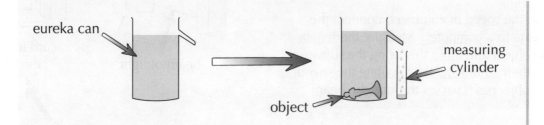

eureka can

measuring cylinder

object

PRACTICAL TIP

Watch out for parallax when taking readings...

Whether you're reading off a ruler, a pipette or a measuring cylinder, make sure you take all readings at <u>eye level</u>. And, if you're taking a reading for a volume, make sure you measure from the bottom of the <u>meniscus</u> (that's from the dip in the curved surface at the top of the liquid).

More on Measuring

Measure **Temperature** Accurately with a **Thermometer**

1) Make sure the <u>bulb</u> of your thermometer is <u>completely submerged</u> in any substance you're measuring the temperature of.

2) Wait for the temperature reading to <u>stabilise</u> before you take your initial reading.

3) Again, read your measurement off the <u>scale</u> on a thermometer at <u>eye level</u>.

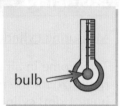

bulb

When you're reading off a scale, use the value of the nearest mark on the scale (the nearest graduation).

You May Have to Measure the **Time Taken** for a Change

1) You should use a <u>stopwatch</u> to <u>time</u> most experiments — they're more <u>accurate</u> than regular watches.

2) Always make sure you <u>start</u> and <u>stop</u> the stopwatch at exactly the right time. Or alternatively, set an <u>alarm</u> on the stopwatch so you know exactly when to stop an experiment or take a reading.

3) You might be able to use a <u>light gate</u> instead (see below). This will <u>reduce the errors</u> in your experiment.

Mass Should Be Measured Using a **Balance**

1) For a <u>solid</u>, set the balance to <u>zero</u> and then place your object onto the scale and read off the mass.

2) If you're measuring the mass of a <u>liquid</u>, start by putting an empty <u>container</u> onto the <u>balance</u>. Next, <u>reset</u> the balance to zero.

3) Then just pour your <u>liquid</u> into the container and record the mass displayed. Easy peasy.

Light Gates Measure **Speed** and **Acceleration**

1) A <u>light gate</u> sends a <u>beam</u> of light from one side of the gate to a <u>detector</u> on the other side. When something passes through the gate, the beam of light is <u>interrupted</u>. The light gate then measures <u>how long</u> the beam was undetected for.

light gate

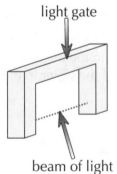

beam of light

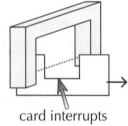

card interrupts the beam

2) To find the <u>speed</u> of an object, connect the light gate to a <u>computer</u>. Measure the <u>length</u> of the object and <u>input</u> this using the software. It will then <u>automatically calculate</u> the speed of the object as it passes through the beam.

3) To measure <u>acceleration</u>, use an object that interrupts the signal <u>twice</u> in a <u>short</u> period of time, e.g. a piece of card with a gap cut into the middle.

Have a look at page 113 for an example of a light gate being used.

4) The light gate measures the speed for each section of the object and uses this to calculate its <u>acceleration</u>. This can then be read from the <u>computer screen</u>.

Working With Electronics

Electrical devices are used in a bunch of experiments, so make sure you know how to use them.

You Have to Interpret **Circuit Diagrams**

Before you get cracking on an experiment involving any kind of electrical devices, you have to plan and build your circuit using a circuit diagram. Make sure you know all of the circuit symbols on page 40 so you're not stumped before you've even started.

You Can Measure **Potential Difference** and **Current**

Voltmeters Measure Potential Difference

1) If you're using an analogue voltmeter, choose the voltmeter with the most appropriate unit (e.g. V or mV).
2) If you're using a digital voltmeter, you'll most likely be able to switch between them.
3) Connect the voltmeter in parallel (p.49) across the component you want to test.
4) The wires that come with a voltmeter are usually red (positive) and black (negative). These go into the red and black coloured ports on the voltmeter. Funnily enough.
5) Then simply read the potential difference from the scale (or from the screen if it's digital).

Ammeters Measure Current

1) Just like with voltmeters, choose the ammeter with the most appropriate unit.
2) Connect the ammeter in series (p.47) with the component you want to test, making sure they're both on the same branch. Again, they usually have red and black ports to show you where to connect your wires.
3) Read off the current shown on the scale or by the screen.

Turn your circuit off between readings to prevent wires overheating and affecting your results (p.41).

Multimeters Measure Both

1) Instead of having a separate ammeter and voltmeter, many circuits use multimeters. These are devices that measure a range of properties — usually potential difference, current and resistance.
2) If you want to find potential difference, make sure the red wire is plugged into the port that has a 'V' (for volts).
3) To find the current, use the port labelled 'A' or 'mA' (for amps).
4) The dial on the multimeter should then be turned to the relevant section, e.g. to 'A' to measure current in amps. The screen will display the value you're measuring.

Don't get your wires in a tangle when you're using circuits...

When you're dealing with voltmeters, ammeters and multimeters, you need to make sure that you wire them into your circuit correctly, otherwise you could mess up your readings. Just remember, the red wires should go into the red ports and the black wires should go into the black ports.

Safety and Experiments

There's <u>danger</u> all around, particularly in science experiments. But don't let this put you off. Just be aware of the <u>hazards</u> and take <u>sensible precautions</u>. Read on to find out more...

Be **Careful** When You Do Experiments

1) There are always hazards in any experiment, so <u>before</u> you start an experiment you should read and follow any <u>safety precautions</u> to do with your method or the apparatus you're using.

2) Stop masses and equipment falling by using <u>clamp stands</u>.

3) Make sure any masses you're using in investigations are of a <u>sensible weight</u> so they don't break the equipment they're used with. Also, make sure strings used in <u>pulley systems</u> are of a sensible <u>length</u>. That way, any hanging masses won't <u>hit the floor</u> or the <u>table</u> during the experiment.

4) When <u>heating</u> materials, make sure to let them <u>cool</u> before moving them, or wear <u>insulated gloves</u> while handling them. If you're using an <u>immersion heater</u> to heat liquids, you should always let it <u>dry out</u> in air, just in case any liquid has leaked inside the heater.

5) If you're using a <u>laser</u>, there are a few safety rules you must follow. Always wear <u>laser safety goggles</u> and never <u>look directly into</u> the laser or shine it <u>towards another person</u>. Make sure you turn the laser <u>off</u> if it's not needed to avoid any accidents.

6) When working with electronics, make sure you use a <u>low</u> enough <u>voltage</u> and <u>current</u> to prevent wires <u>overheating</u> (and potentially melting) and also to avoid <u>damaging components</u>, e.g. blowing a filament bulb.

7) You also need to be aware of <u>general safety</u> in the lab — handle <u>glassware</u> carefully so it doesn't <u>break</u>, don't stick your fingers in sockets and avoid touching frayed wires. That kind of thing.

BEWARE — hazardous physics experiments about...

Before you carry out an experiment, it's important to consider all of the <u>hazards</u>. Hazards can be anything from <u>lasers</u> to <u>electrical currents</u>, or weights to heating equipment. Whatever the hazards, make sure you know all the <u>safety precautions</u> you should follow to keep yourself <u>safe</u>.

Practice Exams

Once you've been through all the questions in this book, you should feel pretty confident about the exams. As final preparation, here is a set of **practice exams** to really get you set for the real thing. The time allowed for each paper is 1 hour 45 minutes. These papers are designed to give you the best possible preparation for your exams.

CGP Practice Exam Paper
GCSE Physics

GCSE Physics

Paper 1

Higher Tier

In addition to this paper you should have:
• A ruler.
• A calculator.
• The Physics Equation Sheet (on the inside back cover).

Centre name				
Centre number				
Candidate number				

Time allowed:
• 1 hour 45 minutes

Surname	
Other names	
Candidate signature	

Instructions to candidates
• Write your name and other details in the spaces provided above.
• Answer **all** questions in the spaces provided.
• Do all rough work on the paper.
• Cross out any work you do not want to be marked.

Information for candidates
• The marks available are given in brackets at the end of each question.
• There are 100 marks available for this paper.
• You are allowed to use a calculator.
• You should use good English and present your answers in a clear and organised way.
• For Questions 3 and 12, ensure that your answers have a clear and logical structure, include the right scientific terms, spelt correctly and include detailed, relevant information.

Advice to candidates
• In calculations show clearly how you worked out your answers.

	For examiner's use						
Q	Attempt Nº			Q	Attempt Nº		
	1	2	3		1	2	3
1				8			
2				9			
3				10			
4				11			
5				12			
6				13			
7							
			Total				

1 A representation of the particles of a substance is shown in **Figure 1**.

Figure 1

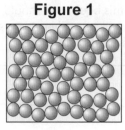

1.1 Name the state of matter of the substance in **Figure 1**.

liquid

[1 mark]

1.2 The substance in **Figure 1** is heated.
Describe what happens to the internal energy of the substance when it is heated.
Explain why this occurs.

*the internal energy would increase
more thermal energy is being
transferred*

[2 marks]

A student is carrying out an experiment into changes of state.

1.3 The student heats a sample of solid wax until it becomes a liquid.
Name this change of state.

Melting

[1 mark]

The student wishes to find the density of the solid wax.
A sample of solid wax has a mass of 0.36 kg, and a volume of 4.0×10^{-4} m³.

1.4 Write down the equation that links mass, volume and density.

density = mass ÷ volume

[1 mark]

1.5 Calculate the density of the wax.

density = 0.36 × 4.0×10⁻⁴ = 0.060144

Density = *0.060144* kg/m³

[2 marks]

2   A student wanted to model how the thickness of an insulating layer affects how quickly the contents of a hot water tank cools. To model this system in the lab, she carried out an investigation to test how the thickness of a cotton wool jacket affected the rate of cooling of a beaker of hot water.
The apparatus she used is shown in **Figure 2**.

Figure 2

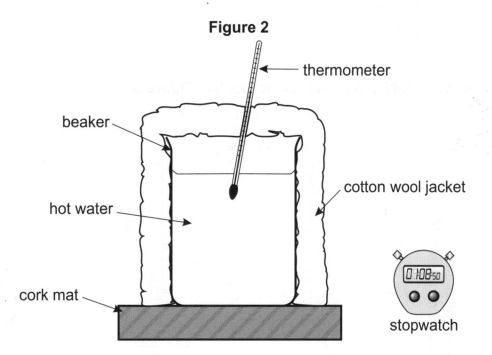

2.1   Give **one** strength and **one** limitation of using the apparatus shown in **Figure 2** to model a hot water tank.

Strength = *the cork mat helps insulate the hot water*

Limitation = *the thermometer is not sized, this may affect accuracy*

[2 marks]

2.2   State the **independent variable** in this investigation.

thickness of the wool

[1 mark]

2.3   State **one** control variable in this investigation.

Volume of hot water

[1 mark]

Question 2 continues on the next page

Turn over ▶

2.4 When the student repeated the investigation, she got very similar results.
Which **one** of the following describes her results?
Tick **one** box.

☑ precise

☐ systematic

☐ anomalous

[1 mark]

The student's results are shown on the graph in **Figure 3**.

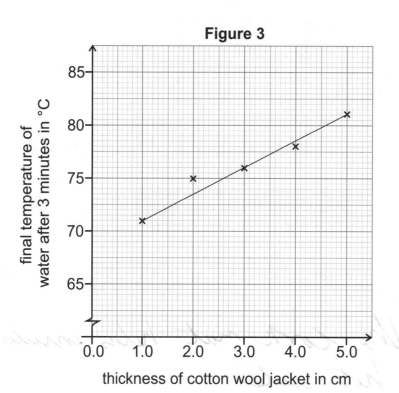

Figure 3

2.5 What conclusion can be made from the results in **Figure 3**?

there is a positive correlation
between the thickness of the wool and temperature

[1 mark]

3   The model of the atom has developed over time.

Figure 4 shows an early model of the atom, Model X, and a currently used model of the atom, Model Y.

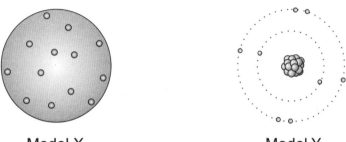

Model X   Model Y

Explain how experiments and scientific discoveries caused our understanding of the atom to develop from Model X to Model Y.

Your answer should include descriptions of the models of the atom shown in **Figure 4**.

....

....

....

....

....

....

....

....

....

....

....
[6 marks]

Turn over for the next question

Turn over ▶

4   **Figure 5** shows a graph of temperature against time for the substance as it is being continually heated.

Figure 5

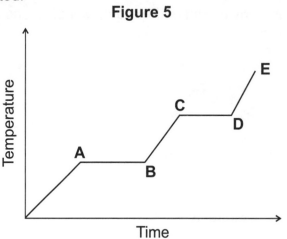

4.1   Identify the state of matter of the substance at point **E**.

.....Gas.....

[1 mark]

4.2   Which point represents the **boiling point**?
Tick **one** box.

☐   A

☐   B

☑   C

[1 mark]

4.3   Explain the shape of the line in **Figure 5** between points **A** and **B**.

The line plateaus because energy is being used to overcome the intermolecular forces of the substance

[2 marks]

4.4   69 000 J of energy is transferred to 1.5 kg of the substance to completely melt it without changing its temperature.
Calculate the specific latent heat of fusion of the substance.
Use the correct equation from the Physics Equation Sheet on the inside back cover.

$$E = mL \qquad L = \frac{69000}{1.6} = 46,000$$

$$L = \frac{E}{M}$$

Specific latent heat of fusion = ...46,000..... J/kg

[3 marks]

5 A scientist is experimenting with static electricity. She combs her hair fifty times with the same comb. Afterwards, some of her hairs stand on end away from each other. The comb becomes negatively charged with a charge of –0.2 nC.

Figure 6

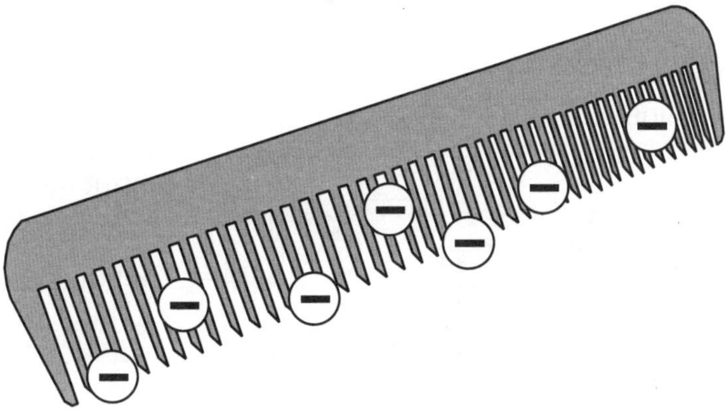

5.1 Explain why the comb has become negatively charged.

Electron have been transferred to the comb

[1 mark]

5.2 Explain why some of the scientist's hairs stand on end away from each other.

AS the hair has transfered electron they are now positively charged and repel each other

[2 marks]

5.3 Give the total charge on the scientist's hair.

Charge = ...0.2......... nC

[1 mark]

5.4 What will happen if the scientist brings the comb close to her hair?
Tick **one** box.

☑ It will attract the hair.

☐ It will repel the hair.

☐ No interaction will be observed.

[1 mark]

Turn over for the next question

Turn over ▶

6 A student carried out an investigation into the specific heat capacity of liquids using the apparatus shown in **Figure 7**. He used identical electric heating coils to heat a beaker of water and a beaker of oil. He used exactly 1 kg of each liquid.

Figure 7

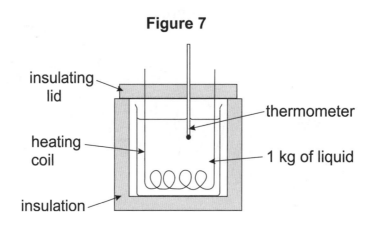

The student recorded the temperature of both the liquids before heating, and then again after ten minutes of heating. His results are shown in **Table 1**.

Table 1

	Water	Oil
Initial temperature in °C	18	18
Final temperature in °C	48	93

6.1 State what is meant by 'specific heat capacity'.

the amount of energy required to raise the temperature of 1kg substance by 1°C

[1 mark]

6.2 During the experiment, the heating coil transferred 126 kJ of energy to each liquid. Use the data from the experiment to calculate the specific heat capacity of the oil. Use the correct equation from the Physics Equation Sheet on the inside back cover.

$$\Delta E = m \, c \, \Delta \theta$$

$$c = \frac{\Delta E}{m \times \Delta \theta} = \frac{126\,000}{1 \times 75}$$

$$= 1680$$

Specific heat capacity of oil = ...1680.... J/kg°C

[4 marks]

6.3 Both oil and water can be used in heating systems.
With reference to the specific heat capacities of oil and water, explain why most heating systems use water rather than oil.

....

....

....

[2 marks]

Turn over for the next question

7   **Table 2** gives details of some isotopes.

Table 2

Isotope	Symbol	Type of decay
Radium-226	$^{226}_{88}\text{Ra}$	alpha
Radon-222	$^{222}_{86}\text{Rn}$	alpha
Radon-224	$^{224}_{86}\text{Rn}$	beta
Bismuth-210	$^{210}_{83}\text{Bi}$	alpha, beta
Bismuth-214	$^{214}_{83}\text{Bi}$	alpha, beta
Lead-210	$^{210}_{82}\text{Pb}$	beta

7.1   Calculate the number of neutrons in a bismuth-214 nucleus.

....

                                                                                        [1 mark]

7.2   Using data from **Table 2**, complete the equations in **Figure 8** to show how the
        following isotopes decay.

Figure 8

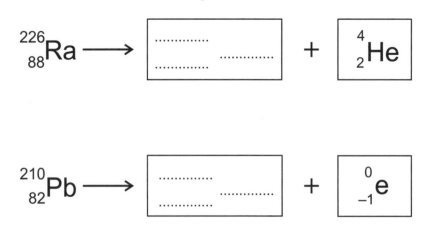

                                                                                        [2 marks]

Figure 9 shows the activity-time graph of a sample of polonium-210.

Figure 9

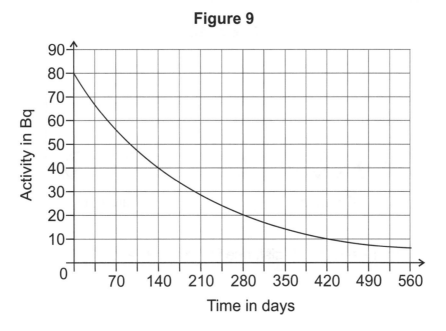

7.3 Using the graph in **Figure 9**, determine the time it takes for the activity of the sample to drop from 80 Bq to 10 Bq.

Time taken = ... days

[1 mark]

7.4 Determine the half-life of polonium-210.

Half-life = ... days

[1 mark]

7.5 Polonium-210 emits alpha radiation.
Ingesting any amount of an alpha source can be very harmful.
Explain why sources of alpha radiation are much more dangerous inside the body than outside the body.

....

....

....

....

....

....

....

....

[4 marks]

Turn over for the next question

Turn over ▶

8   A student is investigating the two electrical circuits shown in **Figure 10**.
All the lamps and batteries used are identical.

Figure 10

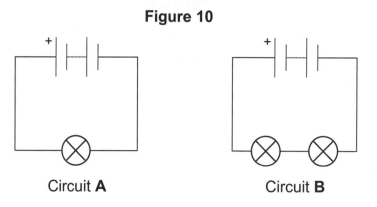

Circuit **A**   Circuit **B**

8.1   Compare the current and total resistance in circuits **A** and **B**.

....

....

....

[2 marks]

8.2   The student adds an ammeter and a voltmeter to circuit **A**.
They show readings of 0.30 A and 11 V respectively.
State the equation that links power, current and potential difference.

....

[1 mark]

8.3   Calculate the power of the lamp.

....

....

....

Power = ... W

[2 marks]

The student considers adding one of the components in **Figure 11** to circuit **A**.

Figure 11

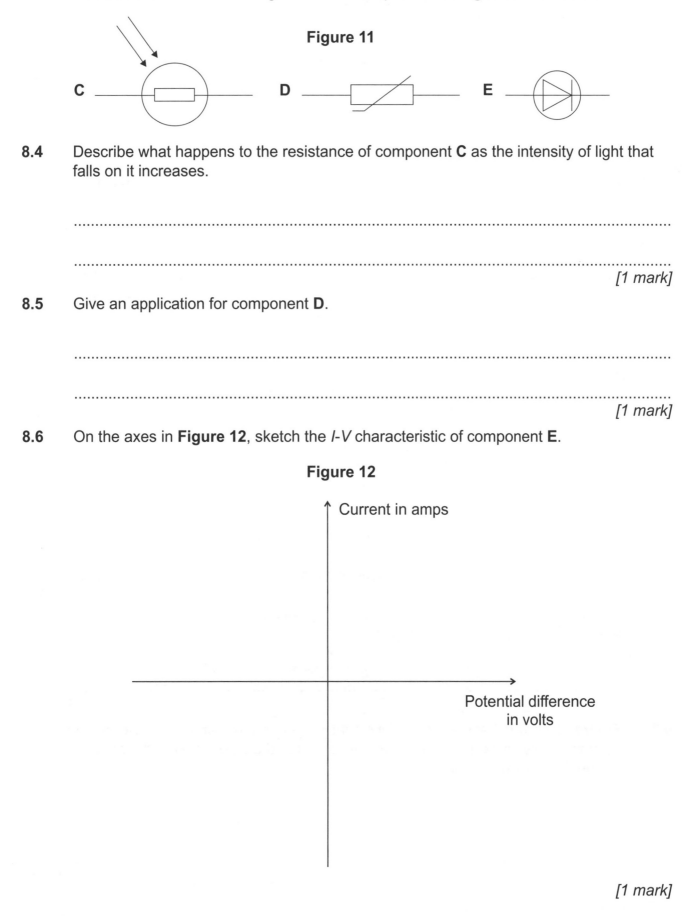

8.4 Describe what happens to the resistance of component **C** as the intensity of light that falls on it increases.

....

....
[1 mark]

8.5 Give an application for component **D**.

....

....
[1 mark]

8.6 On the axes in **Figure 12**, sketch the *I-V* characteristic of component **E**.

Figure 12

[1 mark]

Turn over for the next question

Turn over ▶

9 **Figure 13** shows the amount of electricity generated by different renewable energy resources in the UK each season between 2012 and 2015.
A kilowatt hour (kWh) is a unit of energy equal to 3.6×10^6 J.

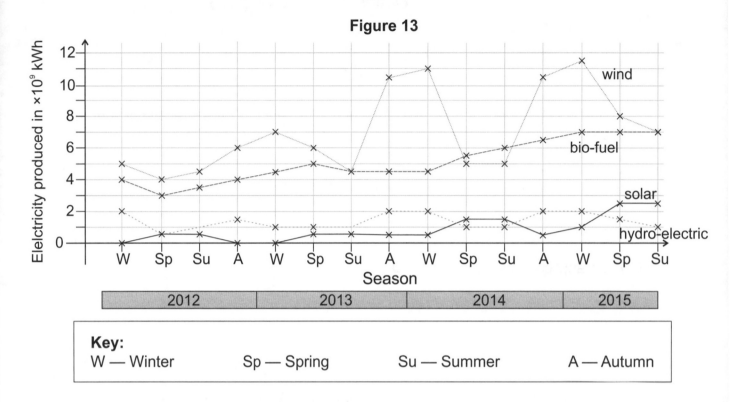

Figure 13

9.1 Using **Figure 13**, determine the amount of electricity generated by bio-fuels and by hydro-electric power in summer 2014.

Bio-fuels = ... kWh

Hydro-electric = ... kWh
[2 marks]

9.2 Using **Figure 13**, suggest which renewable energy resource usually provides the largest amount of electricity to the UK.

....

[1 mark]

9.3 **Figure 13** shows that the amount of electricity generated from solar power during summer is always larger than the amount generated during winter of the same year. Suggest a reason for this.

....

....

[1 mark]

9.4 The majority of electricity in the UK is generated from non-renewable energy resources. Give **one** advantage and **one** disadvantage of using non-renewable energy resources to generate electricity.

Advantage = ...

....

Disadvantage = ...

....

[2 marks]

9.5 Give **two** limitations to increasing the amount of electricity produced from renewable energy resources in the UK.

....

....

....

....

[2 marks]

9.6 Another renewable energy resource is geothermal power.
Suggest why geothermal power is **not** widely used in the UK.

....

....

[1 mark]

Turn over for the next question

Turn over ▶

10 A student wanted to know how the current flowing through a filament lamp changes with the potential difference across it. She set up the circuit shown in **Figure 14**.

Figure 14

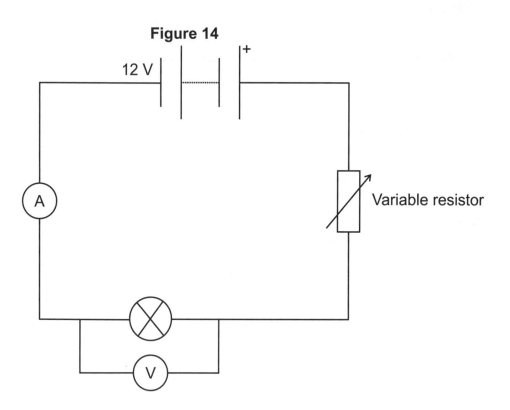

She used the variable resistor to change the potential difference across the lamp. For each setting of the variable resistor, the student recorded the readings of the voltmeter and the ammeter. **Table 3** shows her results.

Table 3

Voltmeter reading in volts	Ammeter reading in amps
0.0	0.00
3.0	0.10
5.0	0.14
7.0	0.17
9.0	0.19
11.0	0.21

Question 10 continues on the next page

10.1 Plot the results displayed in **Table 3** on the grid in **Figure 15**.

Figure 15

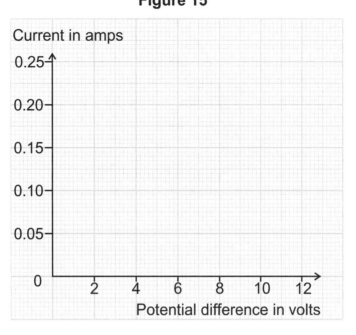

[2 marks]

10.2 Draw a curve of best fit on your graph on the grid in **Figure 15**.

[1 mark]

10.3 Explain, with reference to energy transfers, what happens to the resistance of the lamp as the current through it increases.

....

....

....

....

....

....

[4 marks]

10.4 The lamp is disconnected from the test circuit, and is connected to a 15 V power supply.
At this potential difference the lamp has a resistance of 60 Ω.
The lamp operates at this potential difference for 180 s.
Calculate the amount of charge which passes through the lamp in this time.

....

....

....

....

Charge = ... C

Turn over for the next question

[5 marks]

Turn over ▶

11 **Figure 16** shows the inside of the three-pin plug for an electric kettle.

Figure 16

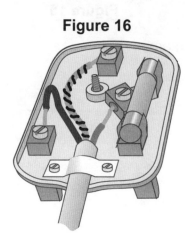

11.1 Give the colours of the earth wire and the live wire.

Earth wire = ...

Live wire = ...

[2 marks]

11.2 Describe the purpose of the earth wire.

....

....

[1 mark]

11.3 Explain how the live wire can be dangerous even when the kettle is switched off.

....

....

....

[2 marks]

11.4 The electric kettle is plugged into an electricity supply and is switched on for 180 s.
While switched on, a current of 10.0 A passes through the kettle.
The resistance of the kettle is 23.0 Ω.
Calculate the energy transferred by the kettle.

....

....

....

....

Energy transferred = ... J

[4 marks]

12 The national grid is a network of pylons and cables that are used to transmit large amounts of electrical power throughout the UK.
The national grid is considered to be an efficient way of transmitting electricity, through its use of devices such as transformers.

Explain how the national grid can transmit high power electricity efficiently, and provide their consumers with useful electric power.

....

....

....

....

....

....

....

....

....

....

....

....

....

[6 marks]

Turn over for the next question

Turn over ▶

13 At the start of a roller coaster ride, a carriage is raised through a vertical height of 20 m to point **A**, as shown in **Figure 17**.

Figure 17

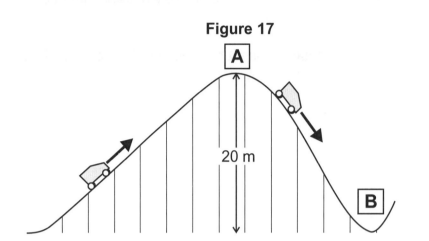

The mass of the empty carriage is 600 kg.
The Earth's gravitational field strength is 9.8 N/kg.

13.1 Write down the equation which links gravitational potential energy, mass, the gravitational field strength and height.

....
[1 mark]

13.2 Calculate the energy transferred (in kJ) to the gravitational potential energy store of the empty carriage as it is lifted to point **A**.
Give your answer to 2 significant figures.

....

....

....

....

Energy transferred to gravitational potential energy store = ... kJ
[2 marks]

During the ride the carriage drops from point **A** to point **B**, as shown in **Figure 17**. **Figure 18** shows a graph of the energy in the carriage's gravitational potential energy store and kinetic energy store as it travels from point **A** to point **B**, assuming there is no friction or air resistance acting on the carriage.

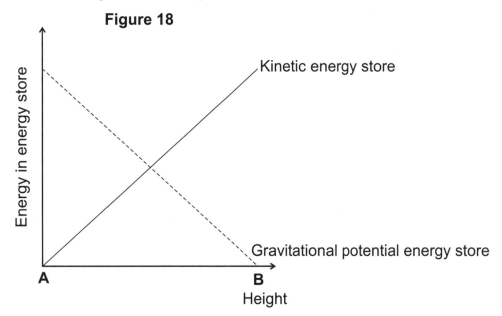

Figure 18

13.3 Explain the shape of the lines in **Figure 18** in terms of energy transfer and conservation of energy.

....

....

....

....

[2 marks]

13.4 Point **A** is 20 m above point **B**.
Calculate the speed of the empty carriage as it reaches point **B**, assuming there is no friction or air resistance acting on the carriage.
Use the correct equation from Physics Equation Sheet on the inside back cover.
Use your unrounded answer from **13.2**.
Give your final answer to 2 significant figures.

....

....

....

....

Speed of carriage at point B = ... m/s

[4 marks]

END OF QUESTIONS

GCSE Physics

Paper 2

Higher Tier

In addition to this paper you should have:
- A ruler.
- A calculator.
- A protractor.
- The Physics Equation Sheet
  (on the inside back cover).

Centre name				
Centre number				
Candidate number				

Time allowed:
- 1 hour 45 minutes

Surname	
Other names	
Candidate signature	

Instructions to candidates
- Write your name and other details in the spaces provided above.
- Answer **all** questions in the spaces provided.
- Do all rough work on the paper.
- Cross out any work you do not want to be marked.

Information for candidates
- The marks available are given in brackets at the end of each question.
- There are 100 marks available for this paper.
- You are allowed to use a calculator.
- You should use good English and present your answers in a clear and organised way.
- For Questions 5.2 and 12.3, ensure that your answers have a clear and logical structure, include the right scientific terms, spelt correctly and include detailed, relevant information.

Advice to candidates
- In calculations show clearly how you worked out your answers.

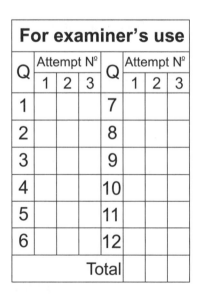

1   X-rays can be used in hospitals for medical treatments and diagnoses.
Figure 1 shows X-ray images used for diagnoses.
X-rays are directed at a body part being examined.
A detector is placed behind the body part to detect the X-rays that reach it.

Figure 1

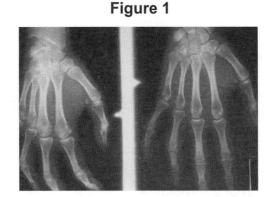

1.1   Explain how X-ray images, like those shown in **Figure 1**, are formed.

X-rays are capable of passing through skin but are blocked by the skeleton

[2 marks]

1.2   Explain why precautions must be taken when using X-rays.

X-rays

[1 mark]

1.3   When radiographers perform X-ray scans in hospitals, they often stand behind a lead
screen while the X-rays are being used.
Suggest why the screen is made of lead.

[1 mark]

1.4   Give **two** medical conditions that X-rays can be used to either treat or diagnose.

[2 marks]

Turn over for the next question

Turn over ▶

2   A remote control uses infrared radiation with a frequency of 2.5×10^{14} Hz to transmit signals to a television.

2.1   State what is meant by the 'frequency' of radiation.

....

[1 mark]

The speed of infrared radiation travelling through air is approximately 3.0×10^8 m/s.

2.2   Write down the equation which links wave speed, frequency and wavelength.

....

[1 mark]

2.3   Calculate the wavelength of the radiation.

....

....

....

Wavelength = ... m

[3 marks]

A student investigated the infrared radiation emitted by hot objects. **Figure 2** shows a Leslie cube, which was filled with hot water. The Leslie cube has four coated faces: shiny black, shiny white, matt white and matt black.

Figure 2

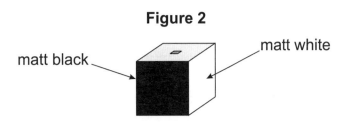

The student placed an infrared detector 10 cm from each face, one at a time, and recorded the intensity of infrared radiation detected. The results are shown on the bar chart in **Figure 3**.

Figure 3

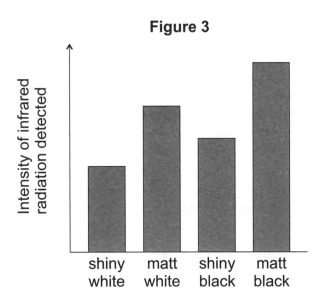

Faces of the Leslie cube

2.4 Infrared emission is affected by both the texture and colour of a surface.
Write down a conclusion from the data in **Figure 3** about how each of these features affect infrared emission.

Texture: ....

....

Colour: ....

....

[2 marks]

2.5 The student repeated her experiment three times. She then used her repeat readings to calculated an average for the value of the intensity of infrared radiation detected at each face.
Give **one** reason why it is important to do this.

....

[1 mark]

Turn over for the next question

Turn over ▶

3 The Sun is a main sequence star.

3.1 Stars are held together by gravitational attraction.
Explain how the Sun remains stable and does not collapse under gravity.

....

....

....

[2 marks]

3.2 The Sun will eventually become a red giant star.
State what the Sun will become immediately after its red giant stage.

....

[1 mark]

The Earth orbits the Sun. The planet Venus is sometimes called Earth's twin.
It has a very similar mass and diameter to the Earth, but is much hotter.
Table 1 shows some data about Venus and Earth.

Table 1

	Mass in kg	Diameter in km	Orbital radius in km	Orbital speed in km/s	Gravitational field strength in N/kg
Venus	5×10^{24}	12 000	1.1×10^8	....	9
Earth	6×10^{24}	13 000	1.5×10^8	30	10

3.3 Using the data from **Table 1**, what is the orbital speed of Venus?
Tick **one** box.

 35 km/s   ☐ 30 km/s   ☐ 25 km/s

[1 mark]

3.4 Using the data in **Table 1**, compare the weight of the same object when
on Venus and Earth. Explain your answer.

....

....

....

....

[2 marks]

4 A swimmer swims a length of a 20 m swimming pool in a straight line.
The distance-time graph in **Figure 4** shows her motion.

Figure 4

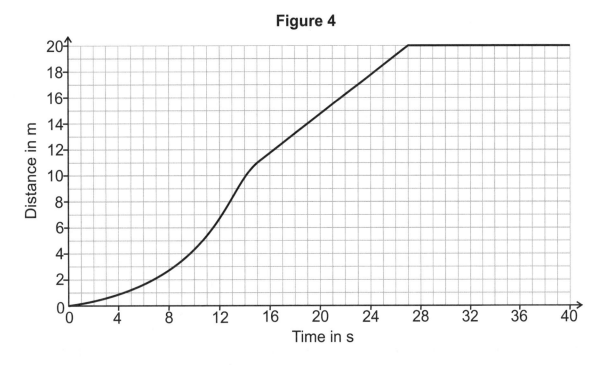

4.1 Calculate the time it takes for the swimmer to complete the length.

Time = ... s
[1 mark]

4.2 For part of her swim, the swimmer is travelling at a constant speed.
Work out the time she spends travelling at a constant speed.

....

Time = ... s
[1 mark]

4.3 Determine the resultant force on the swimmer when she is travelling
at a constant speed.

Force = ... N
[1 mark]

4.4 Between which of the following distances was the swimmer travelling fastest?
Tick **one** box.

☐ Between 14 m and 15 m.

☐ Between 9 m and 10 m.

☐ Between 0 m and 1 m.

[1 mark]

Question 4 continues on the next page

Turn over ▶

A camera travels along the length of the pool to film the swimmer's length.
It travels at a constant speed, and reaches the end of the pool in 25 s.

4.5   Write down the equation which links speed, distance and time.

....

[1 mark]

4.6   Calculate the speed of the camera.

....

....

....

Speed = ... m/s

[3 marks]

4.7   On **Figure 4**, draw a distance-time graph to represent the motion of the camera.

[2 marks]

4.8   The camera cannot film the swimmer if it is behind her.
Using **Figure 4**, explain whether the camera will be able to film the swimmer for the whole length.

....

....

....

....

[2 marks]

5 A student is given a set of apparatus, set up as shown in **Figure 5**.

Figure 5

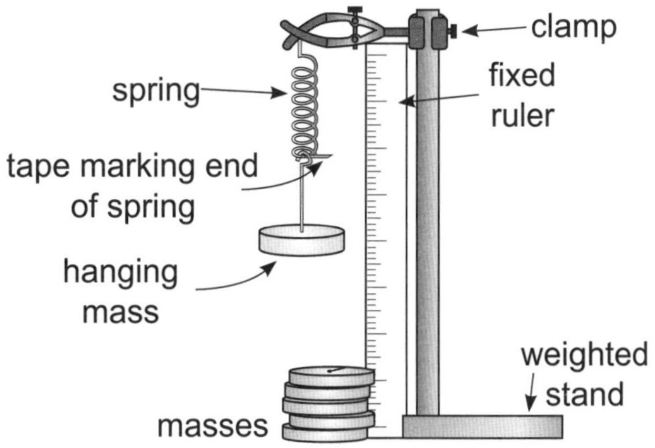

5.1 Name the type of error which may be reduced by the use of the tape marker at the end of the spring.

....
[1 mark]

5.2 Describe how the student could use the experimental set-up in **Figure 5** to plot a force-extension graph for the spring and find the spring constant.

....

....

....

....

....

....

....

....

....

....

....

Question 5 continues on the next page   *[6 marks]*

Turn over ▶

The student used the apparatus in **Figure 5** to produce the graph shown in **Figure 6**.

Figure 6

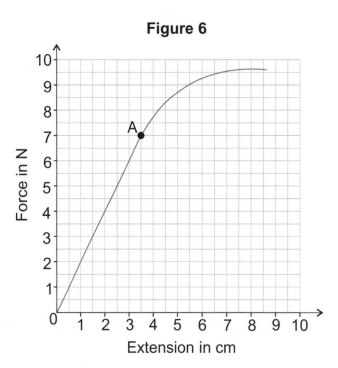

5.3 The student marks point **A** on her graph where the shape of the graph changes. Name point **A**.

....

....
[1 mark]

5.4 Using **Figure 6**, calculate the spring constant of the spring used in the experiment.

....

....

....

....

....

Spring constant = ... N/m
[3 marks]

6 A student made a simple transformer from an iron core and two lengths of wire, as shown in **Figure 7**. He connected a 12 V alternating power supply to one of the coils and a lamp and voltmeter to the other coil.

Figure 7

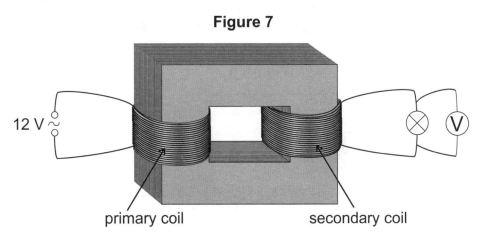

primary coil   secondary coil

6.1 Explain how a potential difference is generated in the secondary coil.

....

....

....

....

[2 marks]

The student experimented with the transformer by changing the number of turns on the secondary coil and measuring the potential difference across the lamp each time. His results are plotted on **Figure 8**.

Figure 8

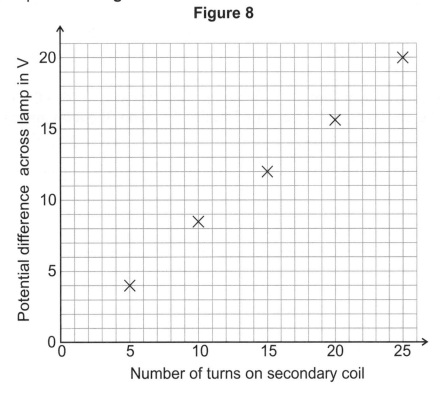

Question 6 continues on the next page

Turn over ▶

6.2   Draw a line of best fit on **Figure 8**.

[1 mark]

The generator in a power station produces an alternating voltage of 25 kV.
This is changed to 400 kV by the transformer shown in **Figure 9**.

Figure 9

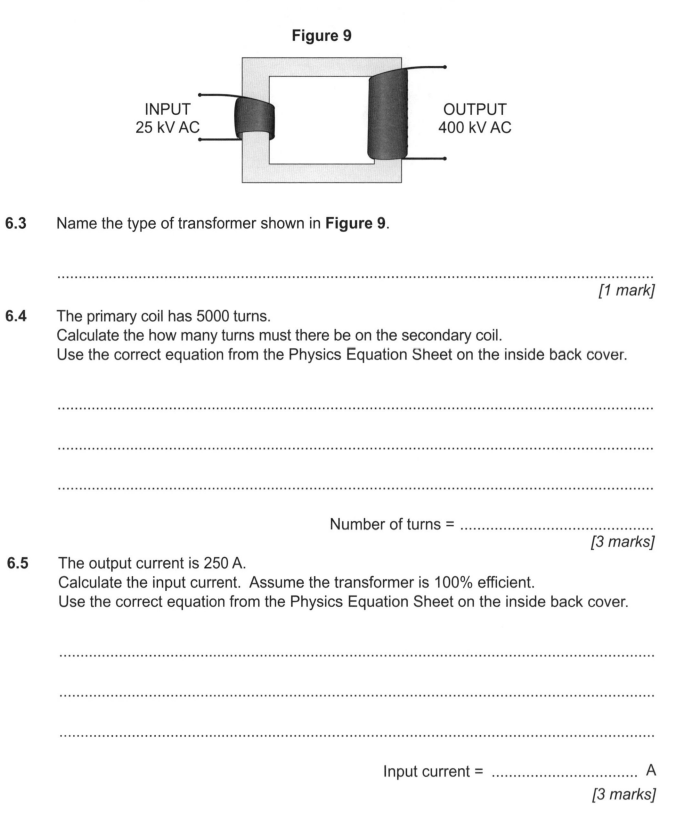

INPUT
25 kV AC

OUTPUT
400 kV AC

6.3   Name the type of transformer shown in **Figure 9**.

....
[1 mark]

6.4   The primary coil has 5000 turns.
Calculate the how many turns must there be on the secondary coil.
Use the correct equation from the Physics Equation Sheet on the inside back cover.

....

....

....

Number of turns = ...
[3 marks]

6.5   The output current is 250 A.
Calculate the input current. Assume the transformer is 100% efficient.
Use the correct equation from the Physics Equation Sheet on the inside back cover.

....

....

....

Input current = ... A
[3 marks]

7 A student wants to investigate the reflection of light from different surfaces. They use a ray box to shine a ray of white light onto a block of material, as shown in **Figure 10**.

Figure 10

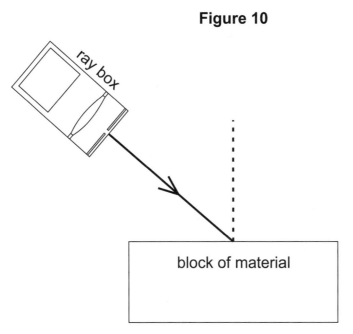

7.1 Measure the angle of incidence of the light ray in **Figure 10**.

Angle of incidence = ... °

[1 mark]

7.2 The student finds that when the light hits the block in **Figure 10**, it undergoes specular reflection. Complete the diagram in **Figure 10** by drawing in the reflected ray.

[2 marks]

The student replaces the block of material with another material. This block of material is transparent. When the ray hits this material, it is transmitted into the block.

7.3 When the light passes into the block, it is bent towards the normal.
Describe how the speed of the light in the block differs from its speed in air.

....

....

[1 mark]

Turn over for the next question

Turn over ▶

8   A driving instructor is looking at the Highway Code. He finds the data shown in **Table 3** about stopping distances for a well-maintained car travelling on dry roads at various speeds.

Table 3

Speed (km/h)	Thinking distance (m)	Braking distance (m)	Stopping distance (m)
32	6	6	12
48	9	14	23
64	12	24	36
80	15	38	53
96	18	55	73
112	21	75	96

8.1   The data in **Table 3** was obtained by observing a large number of drivers. Explain why it was sensible to use a large sample of people.

....

....

....
                                                                                                [2 marks]

8.2   Give **one** factor, other than speed, which affects the braking distance of a car.

....

....
                                                                                                [1 mark]

8.3   A car is travelling at 30 m/s and makes an emergency stop to avoid hitting a hazard. The driver applies the brakes when he is 100 m away from the hazard. Calculate the minimum deceleration required for the car to stop before hitting the hazard. Use the correct equation from the Physics Equation Sheet on the inside back cover.

....

....

....

....

Deceleration = ... m/s^2
                                                                                                [3 marks]

Modern cars contain many safety features. **Figure 11** shows how a collapsible steering column works. When a crash occurs, the driver hits the steering wheel and the steering column collapses.

Figure 11

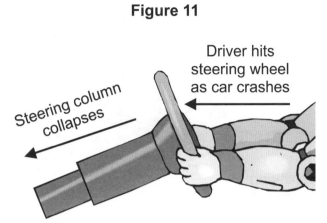

8.4 By considering rate of change of momentum, suggest how a collapsible steering column helps protect the driver from injury in the event of a car crash.

....

....

....

....

[3 marks]

Turn over for the next question

Turn over ▶

9   **Figure 12** shows how a fan can be powered by an electric motor.
When current flows in the wire coil, the fan blades rotate.

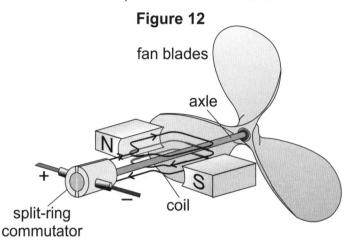

Figure 12

9.1   Name the effect used to rotate the fan.

....

[1 mark]

9.2   Explain how the effect named in **9.1** causes the fan blades to rotate when current flows
in the wire coil.

....

....

....

....

....

....

[2 marks]

The apparatus shown in **Figure 13** can be used to show the force acting on a current-carrying bar in a magnetic field. When the switch is closed, current flows through the metal bar. The metal bar is free to move.

Figure 13

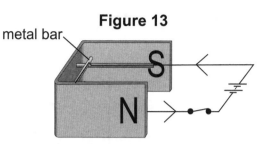

9.3 Explain whether or not a force will act on the bar when the switch is closed. Give a reason for your answer.

....

....

[2 marks]

Microphones also use magnets to work. **Figure 14** shows a diagram of the internal workings of a microphone.

Figure 14

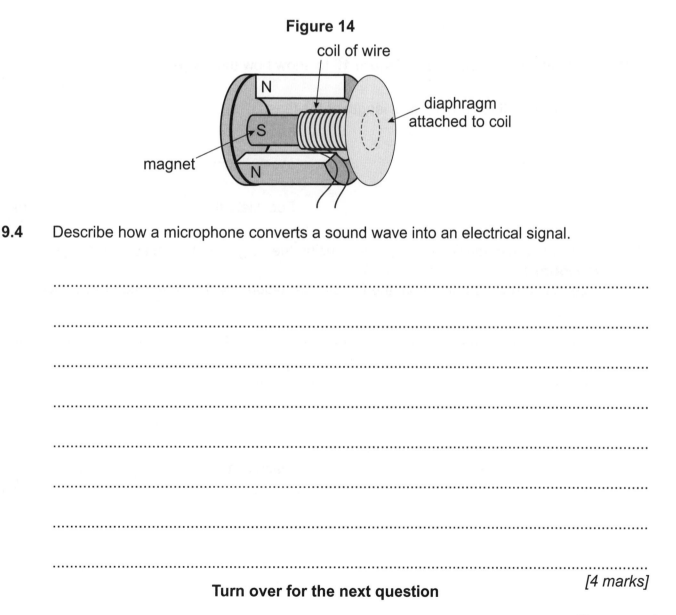

9.4 Describe how a microphone converts a sound wave into an electrical signal.

....

....

....

....

....

....

....

Turn over for the next question

[4 marks]

Turn over ▶

10 A student is looking at an object through a lens.
A diagram of the object, image and lens is shown in **Figure 15**.

Figure 15

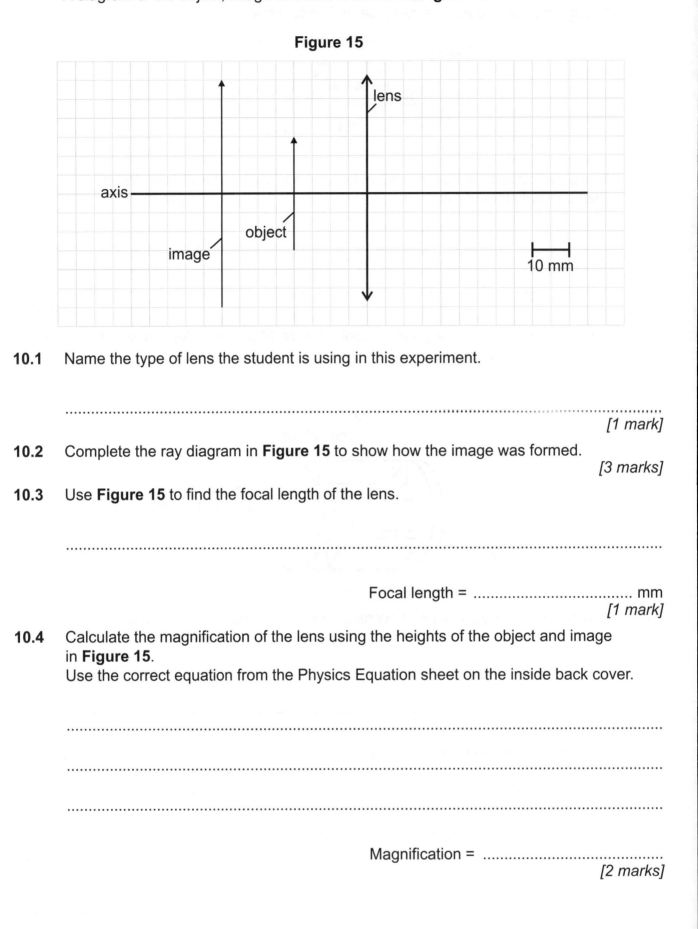

10.1 Name the type of lens the student is using in this experiment.

....
[1 mark]

10.2 Complete the ray diagram in **Figure 15** to show how the image was formed.
[3 marks]

10.3 Use **Figure 15** to find the focal length of the lens.

....

Focal length = ... mm
[1 mark]

10.4 Calculate the magnification of the lens using the heights of the object and image
in **Figure 15**.
Use the correct equation from the Physics Equation sheet on the inside back cover.

....

....

....

Magnification = ...
[2 marks]

11 A child is learning to ride a bicycle. He pushes down on the pedal with a force of 100 N, as shown in **Figure 17**.

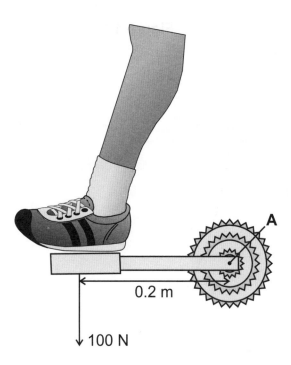

11.1 Write down the equation that links the moment, force, and the distance from the pivot.

....

[1 mark]

11.2 Calculate the moment, about point **A**, that the child exerts, and give the unit.

....

....

....

Moment = ... Units = ...

[3 marks]

11.3 Explain what will happen to the moment about point **A** if the child continues to push on the pedal with the same force while the pedal moves downwards.

....

....

....

....

[2 marks]

Turn over for the next question

Turn over ▶

12 A skydiver jumps from an aeroplane and his motion is recorded.
Figure 16 shows the velocity-time graph of his fall.

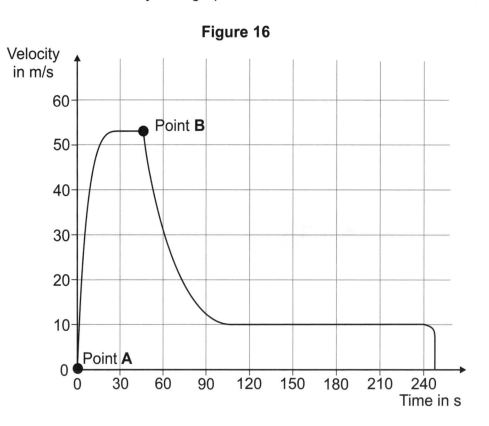

Figure 16

12.1 Write down the equation that links weight, mass and the gravitational field strength.

....

....
[1 mark]

12.2 The skydiver has a mass of 80.0 kg. Calculate his weight.
Use gravitational field strength = 9.8 N/kg.

....

....

Weight = ... N
[2 marks]

12.3 Explain the shape of the graph in **Figure 16** from point **A** to point **B**, in terms of the motion of the skydiver and the forces acting on him.

....

....

....

....

....

....

....

....

....

....

....

....

....

[6 marks]

12.4 Using **Figure 16**, estimate the total distance travelled by the skydiver.

....

....

....

....

Distance = ... m

[3 marks]

END OF QUESTIONS

Topic 1 — Energy

Page 22

Warm-Up Questions

1) Any two from: e.g. mechanically (by a force doing work) / electrically (work done by when a current flows) / by heating / by radiation.

2) Energy is transferred from the chemical energy store of the person's arm to the kinetic energy stores of the arm and the ball (and the gravitational energy store of the ball). Once the ball is released, energy is transferred from the kinetic energy store of the ball to its gravitational potential energy store.

3) kinetic energy = ½ × mass × (speed)^2 / $E_k = \frac{1}{2}mv^2$

4) A lorry travelling at 60 miles per hour.

Exam Questions

1.1 gravitational potential energy = mass × gravitational field strength × height / $E_p = mgh$ *[1 mark]*
So, $h = E_p \div (m \times g)$ *[1 mark]*
$= 137.2 \div (20 \times 9.8)$ *[1 mark]*
$= \mathbf{0.7\ m}$ *[1 mark]*

1.2 Energy is transferred from the gravitational potential energy store *[1 mark]* to the kinetic energy store of the load *[1 mark]*.

1.3 Some of the energy would also be transferred to thermal energy store of the air (and the thermal energy store of the load) *[1 mark]*.

2.1 change in thermal energy = mass × specific heat capacity × temperature change / $\Delta E = mc\Delta\theta$
So, $c = \Delta E \div (m \times \Delta\theta)$ *[1 mark]*
$\Delta\theta = 100 - 20 = 80\ °C$ *[1 mark]*
$c = 36\ 000 \div (0.5 \times 80)$ *[1 mark]*
$= \mathbf{900\ J/kg\ °C}$ *[1 mark]*

2.2 Concrete has a higher specific heat capacity *[1 mark]* and so will be able to store a lot more energy in its thermal energy store *[1 mark]*.

Even if you got the answer to 2.1 wrong, if your conclusion is correct for your answer to 2.1, you'd get the marks for this question.

3.1 Reading values from the start and end of the linear point on the graph:
E.g. $\Delta E = 3 - 1 = 2\ kJ = 2000\ J$
$\Delta\theta = 2.4 - 0.4 = 2\ °C$
[1 mark for any values accurately calculated from a pair of points on the linear part of the graph]
change in thermal energy = mass × specific heat capacity × temperature change / $\Delta E = mc\Delta\theta$,
So, $c = \Delta E \div (m \times \Delta\theta)$ *[1 mark]*
$= 2000 \div (1 \times 2)$ *[1 mark]*
$= \mathbf{1000\ J/kg\ °C}$ *[1 mark]*

3.2 Lower *[1 mark]*. In the investigation, some of the energy transferred by the heater would have been transferred to the thermal energy stores of the surroundings rather than the block *[1 mark]*. For the same temperature change to have occurred for a smaller amount of energy transferred, the specific heat capacity must be smaller *[1 mark]*.

It's important to be able to spot reasons why the results of an investigation aren't perfectly accurate.

Pages 29-30

Warm-Up Questions

1) Energy can be transferred usefully, stored or dissipated, but can never be created or destroyed.

2) Power is the rate of doing work. The units are watts.

3) conduction, convection

4) E.g. lubrication

5) The higher the thermal conductivity, the greater the rate of the energy transfer (i.e. the faster energy is transferred) through it.

6) Some energy is always dissipated, so less than 100% of the input energy transfer is transferred usefully.

Exam Questions

1.1 power = work done ÷ time / $P = W \div t$ *[1 mark]*
$P = 1000 \div 20$ *[1 mark]*
$= \mathbf{50\ W}$ *[1 mark]*

1.2 Less energy is transferred to the thermal energy store of the motor's parts and the surroundings *[1 mark]*, so more is transferred to the scooter's kinetic energy store *[1 mark]*.

1.3 It will be faster / complete the course in less time *[1 mark]* because the motor transfers the same amount of energy, but over a shorter time *[1 mark]*.

2.1 Any two from: e.g. use identical flasks / same temperature gel in each box / same mass of gel in each box / same mass of water in each flask *[1 mark for each correct control variable]*.

2.2 The water at 70 °C *[1 mark]*. The initial gradient of this graph is steeper than the graph for the 60 °C water / there is a greater temperature drop for the 70 °C water over the first minute than for the 60 °C water *[1 mark]*.

2.3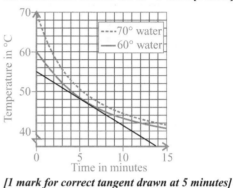
[1 mark for correct tangent drawn at 5 minutes]
E.g. Temperature change = change in $y = 55 - 40 = 15$°C
Change in time = change in $x = 11 - 0 = 11$ mins
Rate of temperature change = $15 \div 11$
$= 1.36... = \mathbf{1.4\ °C/min\ (2\ s.f.)}$
[2 marks for a rate between 1.1 and 1.7 °C/min, otherwise 1 mark for an attempt to calculate gradient of a tangent to curve at 5 minutes]

3.1 $1000 - 280 - 200 = \mathbf{520\ J}$ *[1 mark]*

3.2 efficiency = useful output energy transfer ÷ useful input energy transfer
useful energy = thermal and kinetic energy
$= 200 + 520 = 720\ W$ *[1 mark]*
efficiency = $720 \div 1000$ *[1 mark]*
$= \mathbf{0.72\ (or\ 72\%)}$ *[1 mark]*

3.3 1 minute = 60 seconds, so 4 minutes = $60 \times 4 = 240\ s$
power = energy transferred ÷ time / $P = E \div t$
So, $E = P \div t$ *[1 mark]*
$E = 1000 \times 240$ *[1 mark]*
$= \mathbf{240\ 000\ J}$ *[1 mark]*

4.1 efficiency = useful output energy transfer ÷ useful input energy transfer *[1 mark]*

4.2 efficiency = useful output energy transfer ÷ useful input energy transfer
$= 480 \div 1200$ *[1 mark]*
$= \mathbf{0.4\ (or\ 40\%)}$ *[1 mark]*

4.3 power = energy transferred ÷ time / $P = E \div t$
1 minute = 60 seconds
So, output power = $600 \div 60$ *[1 mark]*
$= \mathbf{10\ W}$ *[1 mark]*

4.4 efficiency = useful output power ÷ total output power
So, total input power = useful output power ÷ efficiency *[1 mark]*
total input power = $10 \div 0.55$ *[1 mark]*
$= 18.181... = \mathbf{18\ W\ (to\ 2\ s.f.)}$ *[1 mark]*

4.5 Disagree *[1 mark]*. Torch B has a lower input energy transfer than torch A, i.e. it transfers less energy per minute than torch A (as $18 \times 60 = 1080$, and $1080 < 1200$) *[1 mark]*.

Even if you got the answer to 4.4 wrong, if your conclusion is correct for your answer to 4.4, you'd get the marks for this question. You could also answer this question by comparing the input powers of the torches.

5.1 E.g. use cavity wall insulation *[1 mark]*.
5.2 How to grade your answer:
Level 0: There is no relevant information. *[No marks]*
Level 1: There is a brief explanation of an experiment to measure how effective the films are as insulation using the equipment shown. The answer lacks coherency. *[1 to 2 marks]*
Level 2: There is an explanation of an experiment to measure how effective the films are as insulation using the equipment shown. There is some discussion of how to compare the effectiveness of the two films. The answer has some structure. *[3 to 4 marks]*
Level 3: There is a clear and detailed explanation of an experiment to measure how effective the films are as insulation using the equipment shown. There is a clear explanation of how to compare the effectiveness of the two films. The answer is well structured. *[5 to 6 marks]*
Here are some points your answer may include:
Wrap one of the beakers in one type of film and another other in the second type of film.
Leave the third unwrapped. This is the control.
Pour boiling water from a kettle into all three beakers.
Cover the beakers with lids to minimise energy loss from the top between temperature readings.
Insert a thermometer through the lid of each beaker.
Record the starting temperature of the water in each beaker.
Take the temperature of the water after regular intervals of time, e.g. every minute, for ten minutes.
Compare the temperatures of the water in the wrapped beakers with the temperature of the water in the control beaker at each reading.
If the films prevent energy transfer, the temperature of the film-covered beakers should be higher than the control at most of the recorded times.
Compare the water temperatures of the water in the wrapped beakers to each other.
If one film is a better insulator than the other, the temperature of the water in that beaker should be higher than the other for most of the recorded times.
5.3 E.g. see if someone else gets the same result using different equipment *[1 mark]*.

Page 38
Warm-Up Questions
1) Any three from: coal / oil / natural gas / nuclear fuel (plutonium or uranium).
2) Advantage: E.g. they'll never run out.
Disadvantage: Any one from: e.g. energy output often depends on outside factors which cannot be controlled / they can't respond to immediate increases in energy demands.
3) E.g. Bio-fuels can be used to run vehicles / electricity generated using renewable resources can be used to power vehicles.
4) Any two from: e.g. it releases greenhouse gases and contributes to global warming / it causes acid rain / coal mining damages the landscape.
5) Any two from: e.g. renewable resources don't currently provide enough energy / energy from renewables cannot be relied upon currently / it's expensive to build new renewable power plants / it's expensive to switch to cars running on renewable energy.

Exam Questions
1.1 Any two from: e.g. wave / tidal / geothermal / bio-fuels *[1 mark for each correct renewable energy resource]*.
1.2 Any one from: e.g. flooding a valley for a dam destroys animal habitats / carbon dioxide is released by rotting vegetation in flooded valley *[1 mark]*.
2.1 From the gravitational potential energy store of the water *[1 mark]* to its kinetic energy store *[1 mark]*.
2.2 E.g. they cause no pollution / they use a renewable energy source *[1 mark for each correct advantage, up to a maximum of two]*.
3.1 power = energy transferred ÷ time / $P = E \div t$
So, $E = P \times t$ *[1 mark]*
seconds in 5 hours = $5 \times 60 \times 60 = 18\,000$ s
Energy provided by 1 m² solar panel
in 5 hours = $200 \times 18\,000$ *[1 mark]*
$= 3\,600\,000$ J *[1 mark]*
Number of panels needed = energy needed ÷ energy provided
$= 32\,500\,000 \div 3\,600\,000$ *[1 mark]*
$= 9.027... =$ **10 panels (to next whole number)** *[1 mark]*
Remember, because you have to have a set number of whole panels, if you get a decimal answer, you need to round up to the next whole number to be able to provide the right amount of energy.
3.2 Ten 1 m² solar panels are needed, and they have 10 m² of space on their roof (10×1 m² = 10 m²), so the family can install sufficient solar panels *[1 mark]*.
3.3 E.g. Solar panels are less reliable than coal-fired power stations *[1 mark]*. The energy output of the solar panels will vary based on the number of hours of good sunlight, and may not be able to provide enough energy on a given day *[1 mark]*. The energy output of coal-fired power stations is not influenced by environmental factors like weather, and energy output can be increased to meet demand *[1 mark]*.

Topic 2 — Electricity

Page 44
Warm-Up Questions
1) Ohms / Ω
2) The greater the resistance, the smaller the current / the smaller the resistance, the greater the current.
3)
4) Any one from: e.g. a wire / a fixed resistor
5) In parallel.
6) A graph that shows how the current flowing through a component changes as the potential difference across it varies.

Exam Questions
1.1 potential difference = current × resistance / $V = IR$
So, $R = V \div I$ *[1 mark]*
$= 1.5 \div 0.30$ *[1 mark]*
$= $ **5.0 Ω** *[1 mark]*
1.2 charge = current × time / $Q = It$
$= 0.30 \times 35$ *[1 mark]*
$= $ **10.5 C** *[1 mark]*
1.3 The amount of current flowing through the circuit will decrease *[1 mark]*.
1.4
[1 mark]
1.5 The resistance of the filament lamp increases as the temperature increases *[1 mark]*.

2.1   Potential difference (across the component) *[1 mark]*.

2.2   Diodes only allow current to flow in one direction *[1 mark]*.

2.3   At point A, $V = 6$ V, $I = 3$ A *[1 mark]*

       potential difference = current × resistance / $V = IR$

       $R = V \div I$ *[1 mark]*

          $= 6 \div 3$ *[1 mark]*

          $= \mathbf{2\ \Omega}$ *[1 mark]*

Page 51

Warm-Up Questions

1)   Any one from: e.g. in automatic night lights / outdoor lighting / burglar detectors.

2)   The resistance decreases.

3)   E.g.

4)   Any one from: e.g. if you remove or disconnect one component, then the whole circuit is broken / you can't switch components on or off independently.

5)   The total resistance is the sum of all the resistances.

6)   Two resistors connected in series.

7)   E.g.

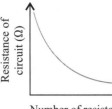

Exam Questions

1.1   total resistance $= R_1 + R_2 + R_3$

          $= 2 + 3 + 5$ *[1 mark]*

          $= \mathbf{10\ \Omega}$ *[1 mark]*

1.2   The current will be 0.4 A *[1 mark]* because in a series circuit, the same current flows through all parts of the circuit *[1 mark]*.

1.3 V = source pd

       $V_3 = V - V_1 - V_2$

          $= 4 - 0.8 - 1.2$ *[1 mark]*

          $= \mathbf{2\ V}$ *[1 mark]*

2.1   15 V *[1 mark]*

Potential difference is the same across each branch in a parallel circuit.

2.2   potential difference = current × resistance / $V = IR$

       $I = V \div R$ *[1 mark]*

          $= 15 \div 3$ *[1 mark]*

          $= \mathbf{5\ A}$ *[1 mark]*

2.3 $I_2 = 5 + 3.75$ *[1 mark]*

          $= \mathbf{8.75\ A}$ *[1 mark]*

Even if you got the answer to 2.2 wrong, award yourself the marks for 2.3 if you did the sum above correctly.

Page 57

Warm-Up Questions

1)   The live wire, neutral wire, and earth wire.

2)   Energy is transferred electrically to the thermal energy store of the heating element.

3)   power = (current)² × resistance / $P = I^2R$

4)   The network of cables and transformers that distributes electricity across the country.

Exam Questions

1   The live wire is at a potential difference of around 230 V *[1 mark]*. Touching the wire forms a low-resistance path from the wire to the earth through your body *[1 mark]*, causing a large current to flow through you, which is an electric shock *[1 mark]*.

2.1   power = potential difference × current / $P = V \times I$ *[1 mark]*

2.2 $I = P \div V$ *[1 mark]*

       $= (2.8 \times 1000) \div 230$ *[1 mark]*

       $= 12.17...$ A $= \mathbf{12}$ *[1 mark]* **A** *[1 mark, allow 'amps']*

2.3   She should choose kettle B because it has the higher power rating *[1 mark]*. This means that it transfers more energy usefully / to heat the water per unit time, so it will boil the water faster *[1 mark]*.

3.1   Step-up transformers increase the potential difference *[1 mark]* of the electricity supply, allowing the electricity to be transmitted at a high potential difference, and so a low current *[1 mark]*. This reduces the energy lost through heating in the cables *[1 mark]*.

3.2   Rearrange $V_sI_s = V_pI_p$ for I_s:

       $I_s = \dfrac{V_pI_p}{V_s}$ *[1 mark]*

          $= \dfrac{4.00 \times 10^5 \times 2.00 \times 10^3}{2.75 \times 10^5}$ *[1 mark]*

       $= 2.909... \times 10^3$ A

       $= \mathbf{2.91 \times 10^3\ A}$ **(to 3 s.f.)** *[1 mark]*

4.1   power = potential difference × current / $P = IV$

       $P = 0.5 \times 3.0$ *[1 mark]*

       $= \mathbf{1.5\ W}$ *[1 mark]*

       energy transferred = power × time / $E = Pt$

       $t = 0.5$ hours $= (30 \times 60)$ s $= 1800$ s

       $E = 1.5 \times 1800$ *[1 mark]*

       $= \mathbf{2700\ J}$ *[1 mark]*

4.2   energy transferred = charge flow × potential difference / $E = QV$

       So, $Q = E \div V$ *[1 mark]*

       $Q = 2700 \div 3.0$ *[1 mark]*

       $= \mathbf{900\ C}$ *[1 mark]*

In the exam, you'd probably get all the marks for 4.2 if you did the sum above correctly, even if you got the answer to 4.1 wrong.

Page 61

Warm-Up Questions

1)   It becomes positively charged.

2)   electrons

3)   attract

4)   electrically charged objects

5)   It increases.

Exam Questions

1.1   The charge on the cloth is +1.6 nC *[1 mark]*.

1.2   Electrons are transferred from the cloth and onto the sphere *[1 mark]*.

1.3   B *[1 mark]*

1.4

    ←——B   ⊖

    [1 mark for arrow drawn at answer to 1.3, pointing away from sphere]

1.5   A strong electric field can ionise particles in the air *[1 mark]*. This makes the air conductive and allows charge to flow to the earth in the form of a spark *[1 mark]*.

2   A *[1 mark]*. The rod is negatively charged so would

Topic 3 — Particle Model of Matter

Pages 69-70
Warm-Up Questions
1) In a liquid, the particles are held close together, but can move past each other. They form irregular arrangements. The particles move in random directions at low speeds.
2) Density is a measure of the amount of mass in a given volume / compactness.
3) Cooling a system decreases its internal energy.
4) The specific latent heat of vaporisation of a substance is the amount of energy needed to change 1 kg of that substance from a liquid to a gas.
5) J/kg
6) Atmospheric pressure decreases with height, so as the helium balloon floats upwards, the pressure outside the balloon decreases. This causes the balloon to expand until the pressure inside drops to the same as the atmospheric pressure.

Exam Questions
1.1 Particles are held close together in a fixed, regular pattern *[1 mark]*. They vibrate about fixed positions *[1 mark]*.
1.2 melting *[1 mark]*
1.3 evaporation *[1 mark]*
2.1 The densities of each of the toy soldiers are the same, but their masses may vary *[1 mark]*.
2.2 The volume of the toy soldier / the volume of water displaced by the toy soldier *[1 mark]*. The mass of the toy soldier *[1 mark]*.
2.3 How to grade your answer:
Level 0: There is no relevant information. *[No marks]*
Level 1: There is a brief explanation of an experiment to measure the density of the toy soldier. The answer lacks coherency. *[1 to 2 marks]*
Level 2: There is an explanation of an experiment to measure the density of the toy soldier, with reference to the equipment needed. The answer has some structure. *[3 to 4 marks]*
Level 3: There is a clear and detailed explanation of an experiment to measure the density of the toy soldier. The method includes details of the equipment needed and how to process the results to work out the density of the toy soldier. The answer is well structured. *[5 to 6 marks]*
Here are some points your answer may include:
Measure and record the mass of the toy soldier using the mass balance.
Fill a eureka can with water.
Place an empty measuring cylinder beneath the spout of the eureka can.
Submerge the toy soldier in the eureka can.
Measure the volume of water displaced from the eureka can using the measuring cylinder.
The volume of water displaced is equal to the volume of the soldier.
Use the equation 'density = mass ÷ volume / $\rho = m \div V$' to calculate the density of the toy soldier.
With questions where you have to describe a method, make sure your description is clear and detailed.
3.1 specific latent heat of fusion *[1 mark]*
3.2 thermal energy for a change of state = mass × specific latent heat / $E = mL$
So, $L = E \div m$ *[1 mark]*
40.8 g = (40.8 ÷ 1000) kg = 0.0408 kg
$L = 47\,100 \div 0.0408$ *[1 mark]*
= **1.154... × 10⁶ J/kg**
= **1.15 × 10⁶ J/kg (to 3 s.f.)** *[1 mark]*

3.3 The internal energy of the system increases as it's heated *[1 mark]*. This is because heating a system transfers energy from thermal energy stores to the kinetic stores of the particles in the system *[1 mark]*.
3.4 When a system is heated, energy is transferred to the particles in the system, causing them to gain energy in their kinetic stores/move faster *[1 mark]*. If the substance is heated enough, the particles will have enough energy in their kinetic stores to break the bonds holding them together, so they change state *[1 mark]*.
4.1 The particles in a gas have high energies *[1 mark]*, move in random directions at high speeds *[1 mark]* and are not arranged in any pattern *[1 mark]*.
4.2 In a sealed container, the gas particles collide with the container walls *[1 mark]* and exert a force on the walls, (creating an outward pressure) *[1 mark]*.
4.3 density = mass ÷ volume / $\rho = m \div V$ *[1 mark]*
4.4 $\rho = 8.2 \div 6.69$ *[1 mark]*
= **1.2 g/cm³** *[3 marks for correct answer, including units. Deduct 1 mark if reported to an incorrect number of significant figures and deduct 1 mark if incorrect units given]*
4.5 The pressure of the gas within the container increases as the container is heated *[1 mark]*. This is because the increase in temperature causes the particles to move faster / increases the energy in the kinetic energy stores of the particles, meaning more force is exerted on the walls of the container when the particles collide with it *[1 mark]*. The increased speed / average energy in kinetic stores of the particles also means there are more collisions, which also increases the total force exerted on the container *[1 mark]*.
5 density = mass ÷ volume / $\rho = m \div V$
So, $m = \rho \times V$ *[1 mark]*
volume of cube = 1.5 × 1.5 × 1.5 = 3.375 cm³ *[1 mark]*
The cube's density is 3500 kg/m³.
1 g/cm³ = 1000 kg/m³, so this is
3500 ÷ 1000 = 3.5 g/cm³ *[1 mark]*
$m = 3.5 \times 3.375$ *[1 mark]*
= 11.8125 = **12 g (to 2 s.f.)** *[1 mark]*
6 pV = constant
So, when $V = 0.034$ m³ and $p = 98$ kPa,
$pV = 0.034 \times 98 = 3.332$ *[1 mark]*
When $V = 0.031$ m³, $p \times 0.031 = 3.332$
So $p = 3.332 \div 0.031$ *[1 mark]*
= **110 kPa (to 2 s.f.)** *[1 mark]*

Topic 4 — Atomic Structure

Pages 79-80
Warm-Up Questions
1) The nuclear model of the atom contains a tiny nucleus which contains protons and neutrons. The rest of the atom is mostly empty space. Electrons exist in fixed energy levels round the outside of the nucleus.
2) Isotopes of an element are atoms with the same number of protons / the same atomic number but a different number of neutrons / a different mass number.
3) alpha particles
4) Beta emitters are not immediately absorbed, like alpha radiation, and do not penetrate as far as gamma rays. Therefore variations in the thickness of the sheet significantly affect the amount of radiation passing through.
5) beta radiation
6) gamma rays
7) Substances with a short half-life decay very quickly so emit high amounts of radiation initially.

Exam Questions

1.1 Beta (particles) *[1 mark]*, because the radiation passes through the paper, but not the aluminium, so it is moderately penetrating in comparison to the other two *[1 mark]*.

1.2 E.g. a Geiger-Muller tube/counter *[1 mark]*

2.1 The time taken for the number of radioactive nuclei in a sample to halve / the time taken for the count-rate or activity to fall to half of its initial level *[1 mark]*.

2.2 4 minutes is equivalent to $4 \div 2 = 2$ half-lives *[1 mark]*. After 1 half-life, there will be ½ of the unstable nuclei left. So, after 2 half-lives, there will be $½ \div 2 = ¼$ / **one quarter** of the unstable nuclei left *[1 mark]*.

3.1 $2 \times 60 = 120$ seconds
$120 \div 40 = 3$ half-lives *[1 mark]*
$8000 \div 2 = 4000$, $4000 \div 2 = 2000$,
$2000 \div 2 = \textbf{1000 Bq}$ *[1 mark]*

3.2 $8000 \div 2 = 4000$, $4000 \div 2 = 2000$, $2000 \div 2 = 1000$,
$1000 \div 2 = 500$, $500 \div 2 = 250$.
So it takes **5 half-lives** to drop to 250 Bq *[2 marks for correct answer, otherwise 1 mark for attempting to halve values to find number of half-lives]*.

3.3 $(100 \div 8000) \times 100$ *[1 mark]* = **1.25%** *[1 mark]*

4.1 Protons *[1 mark]* and neutrons *[1 mark]*.

4.2 The total number of protons and neutrons in the nucleus/atom *[1 mark]*.

4.3 Atom A and atom B *[1 mark]* because isotopes of the same element have the same atomic number, but different mass numbers *[1 mark]*.

4.4 They have opposite charges *[1 mark]*.

You have to use some of your knowledge from Topic 2 here. An electric field will cause an attractive force on one charge, and a repulsive force on the opposite charge, so they'll be deflected in opposite directions.

5.1 E.g. $^{0}_{-1}e$ / $^{0}_{-1}\beta$ *[1 mark]*

5.2 The atomic number increases by 1 *[1 mark]* and the mass number stays the same *[1 mark]*.

5.3 The atomic number doesn't change *[1 mark]* and neither does the mass number *[1 mark]*.

5.4 $^{209}_{84}\text{Po} \rightarrow {}^{205}_{82}\text{Pb} + {}^{4}_{2}\text{He}$
[1 mark for both the mass number and atomic number of He, 1 mark for the atomic number of Po, 1 mark for the mass number of Pb]

Page 85
Warm-Up Questions

1) Any three from: e.g. food / rocks / building materials / cosmic rays / nuclear explosions / nuclear waste.

2) Radioactive contamination is caused when radioactive atoms gets into or onto an object. Radioactive irradiation occurs when objects are exposed to radiation from a radioactive source.

3) Any two from: e.g. always handle a source with tongs / never allow the source to touch the skin / never have the source out of its lead-lined box for longer than necessary / wear gloves when handling the source.

4) E.g. because it is ionising and can damage cells.

5) Alpha radiation can't penetrate the skin and is easily blocked by a small air gap, so cannot damage tissue inside the body through irradiation. If a person is contaminated with alpha radiation however, it is very dangerous as alpha sources are highly ionising and do lots of damage to nearby tissue.

Exam Questions

1.1 The cells will be killed *[1 mark]*.

1.2 Any one from: e.g. so other scientists can check to see if the data is reproducible / so other scientists can check to see if the data is repeatable *[1 mark]*.

2.1 A uranium-235 nucleus absorbs a neutron *[1 mark]* and splits into two smaller nuclei and releases 2 or 3 neutrons *[1 mark]*. Some of these neutrons go on to start other fissions, and so on, creating a chain reaction *[1 mark]*.

2.2 There could be an explosion *[1 mark]*.

3.1 Iodine-123 is absorbed by the thyroid gland just like normal iodine-127, but it is radioactive / gives out radiation *[1 mark]*. This radiation can be detected outside of the patient *[1 mark]*. The radiation produced can be monitored to see if the thyroid gland is working correctly *[1 mark]*.

3.2 Technetium-99m *[1 mark]* because it's got a short half-life, which means it won't be very radioactive inside the patient for long *[1 mark]*. It's also a gamma source, which means it'll pass out of the patient without causing much damage/ionisation *[1 mark]*.

Topic 5 — Forces

Page 92
Warm-Up Questions

1) Contact force: e.g. air resistance
   Non-contact force: e.g. gravitational attraction

2) Vector quantity: e.g. velocity / momentum / force
   Scalar quantity: e.g. mass / speed / volume

3) a) N/kg
   b) kg
   c) N

4) If all force arrows placed tip-to-tail form a closed loop, the forces are balanced.

Exam Questions

1.1 Total force to the right = $1700 + 300 = 2000$ N
Total force to the left = 2000 N
Total horizontal force = $2000 - 2000 = 0$ N *[1 mark]*
Resultant force = downwards force − upwards force
$\qquad = 800 - 300$
$\qquad = \textbf{500 N}$ *[1 mark]* **downwards** *[1 mark]*

1.2 Total vertical force = 0 N
so, $y = \textbf{400 N}$ *[1 mark]*
Total horizontal force = 0 N
so, $x + 500$ N = 2000 N
$\qquad x = 2000 - 500 = \textbf{1500 N}$ *[1 mark]*

2.1 work done = force × distance moved along the line of action of the force / $W = Fs$
$W = 42\,000 \times 700$ *[1 mark]*
$\quad = 29\,400\,000$ J *[1 mark]*
$\quad = \textbf{29 400 kJ}$ *[1 mark]*

2.2

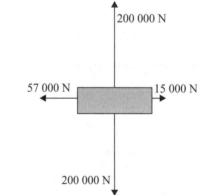

[1 mark for four arrows in directions shown, 1 mark for upwards arrow same length as downwards arrow, 1 mark for driving force arrow bigger than resistive force arrow, 1 mark for all arrows labelled correctly]

You'd still get the marks if you've drawn the driving force to the right, and the resistive force to the left, as the direction that the train was travelling in was not specified in the question.

3.1 The spring would stretch less on Mars because the gravitational field strength on Mars is less than that on Earth *[1 mark]*, so the ball would weigh less *[1 mark]* and the force on the spring would be lower *[1 mark]*.

3.2 weight = mass × gravitational field strength / $W = mg$
so $g = W \div m$ *[1 mark]*
$= 0.37 \div 0.10$ *[1 mark]*
$= \textbf{3.7 N/kg}$ *[1 mark]*

Page 99
Warm-Up Questions
1) Force = spring constant × extension / $F = ke$
2) The limit of proportionality is the point at which the spring stops behaving according to $F = ke$ / F is no longer proportional to e.
3) moment = force × perpendicular distance from pivot / $M = Fd$
4) Nm (newton metres)

Exam Questions
1.1 The mass on the bottom of the spring / the force applied to the bottom of the spring *[1 mark]*.
1.2 Any one from: e.g. the spring used throughout the experiment / the temperature the experiment is carried out at *[1 mark]*.
1.3 extension = 2.5 cm = 0.025 m *[1 mark]*
force = spring constant × extension / $F = ke$
so $k = F \div e$ *[1 mark]*
$= 4 \div 0.025$ *[1 mark]*
$= \textbf{160 N/m}$ *[1 mark]*
Remember to convert the measurement of extension from cm into m before you do your calculation.
1.4 The spring has been inelastically deformed *[1 mark]*.
2.1 distance = 3 cm = 0.03 m *[1 mark]*
moment = force × distance / $M = Fd$
$M = 15 \times 0.03$ *[1 mark]*
$= \textbf{0.45 Nm}$ *[1 mark]*
2.2 distance = 12 cm = 0.12 m *[1 mark]*
$M = Fd$
$= 15 \times 0.12$ *[1 mark]*
$= \textbf{1.8 Nm}$ *[1 mark]*
2.3 The B end should be put into the bolt *[1 mark]* because it produces the greater moment, making the bolt easier to turn when a given force is applied *[1 mark]*.

Page 103
Warm-Up Questions
1) Pressure is the force on a surface per unit area.
2) Height of the column, and the gravitational field strength.
3) 3.5 N
4) As altitude increases, atmospheric pressure decreases.

Exam Questions
1.1 It is more dense than water *[1 mark]*.
1.2 pressure = height of column × density × gravitational field strength / $p = h\rho g$
$h = (50 \div 100) - (8 \div 100) = 0.42$ m *[1 mark]*
$p = 0.42 \times 1000 \times 9.8$ *[1 mark]*
$= 4116$ Pa *[1 mark]*
$= \textbf{4100 Pa (to 2 s.f.)}$ *[1 mark]*
2.1 pressure = force (normal to surface) ÷ area / $p = F \div A$
$p = 175 \div 0.25$ *[1 mark]*
$= \textbf{700 Pa}$ *[1 mark]*
2.2 Since liquid cannot be compressed, pressure on piston 2 is equal to the pressure applied by piston 1.
pressure = force (normal to surface) ÷ area / $p = F \div A$
So, $F = p \times A$ *[1 mark]*
$= 700 \times 1.3$ *[1 mark]*
$= \textbf{910 N}$ *[1 mark]*

Page 110
Warm-Up Questions
1) Speed is scalar, velocity is a vector / velocity has a direction, speed does not.
2) a) E.g. 3 m/s
   b) E.g. 55 m/s
   c) E.g. 250 m/s
Your answers may be slightly different to these, but as long as they're about the same size, you should be fine to use them in the exam.
3) An upwards curved line / a curve with increasing gradient.
4) A straight, horizontal line.
5) As the speed of the car increases, air resistance on the car increases.
6) Her new terminal velocity is lower than her original terminal velocity.

Exam Questions
1.1 The cyclist travels at a constant speed (of 3 m/s) between 5 s and 8 s *[1 mark]*, then decelerates between 8 s and 10 s *[1 mark]*.
1.2 Area of triangle = 0.5 × width × height
Width = 5 − 2 = 3 s
Height = 3 m/s
Distance = 0.5 × 3 × 3 = **4.5 m**
[2 marks, otherwise 1 mark for an attempt to calculate the area under the graph between 2 and 5 seconds]
1.3 Acceleration is given by the gradient of a velocity-time graph.
change in y = 3 − 0 = 3 m/s
change in x = 5 − 2 = 3 s
acceleration = 3 ÷ 3 = **1 m/s²**
[2 marks, otherwise 1 mark for an attempt to calculate the gradient of the line between 2 and 5 seconds]
You could also have used $a = \Delta v \div \Delta t$ here.
1.4 average acceleration = change in velocity ÷ change in time / $a = \Delta v \div \Delta t$
velocity at 8 s = 3 m/s; velocity at 10 s = 2 m/s
so $\Delta v = 2 - 3 = -1$ m/s *[1 mark]*
So, $a = -1 \div 2$ *[1 mark]*
$= -0.5$ m/s²
$= \textbf{0.5 m/s}^2$ *[1 mark]*
Your answer should be positive since the question asks for deceleration, rather than acceleration.
2.1 0 N *[1 mark]*
2.2 Parachutist A is travelling at a constant velocity / their terminal velocity *[1 mark]*.
2.3 700 N *[1 mark]* because in free fall, at terminal velocity the air resistance is equal to the parachutist's weight *[1 mark]*.
2.4 Parachutist A will have a higher terminal velocity *[1 mark]*, since he weighs more *[1 mark]*. Air resistance increases with speed *[1 mark]* so the parachutist will accelerate to a higher speed before air resistance can equal their weight and reach terminal velocity *[1 mark]*.
2.5 When the parachute opens, the surface area of the parachutist increases, so the air resistance increases *[1 mark]* and they decelerate until the forces rebalance and they reach a new, smaller terminal velocity *[1 mark]*.

Page 115
Warm-Up Questions
1) 0 N
2) boulder B
Boulder B needs a greater force to accelerate it by the same amount as boulder A.
3) true
This is Newton's Third Law.
4) Masses should be added to the trolley.

Exam Questions

1.1 The ball exerts a force of –500 N on the bat *[1 mark]*, because, due to Newton's Third Law, if the bat exerts a force on the ball, the ball exerts an equal force on the bat in the opposite direction *[1 mark]*.

1.2 The acceleration of the ball is greater *[1 mark]* because it has a smaller mass, but is acted on by the same size force (and $F = ma$) *[1 mark]*.

2.1 force = mass × acceleration / $F = ma$
So, $a = F \div m$ *[1 mark]*
Set direction of van's motion to be positive, so $F = -200$ N
$a = -200 \div 2500$ *[1 mark]*
$= -0.08$ m/s^2
So, deceleration = **0.08 m/s^2** *[1 mark]*

The question asked for deceleration, so you should really quote your answer without the minus sign. However, you should get the marks either way.

2.2 force = mass × acceleration / $F = ma$
$F = 10.0 \times 29.0$ *[1 mark]*
$= $ **290 N** *[1 mark]*

2.3 By Newton's Third Law, force on van in collision is –290 N *[1 mark]*.
force = mass × acceleration / $F = ma$
So, $a = F \div m$
$a = -290 \div 2500$ *[1 mark]*
$= -0.116$ m/s^2
So deceleration = **0.116 m/s^2** *[1 mark]*

You'd still get the marks here, even if you got 2.2 wrong, as long as your method's correct.

3* How to grade your answer:
Level 0: There is no relevant information. *[No marks]*
Level 1: A simple experiment to investigate force and acceleration which can be performed with the given equipment is partly outlined. The answer lacks coherency. *[1 to 2 marks]*
Level 2: An experiment to investigate force and acceleration which can be performed with the given equipment is outlined in some detail. The answer has some structure. *[3 to 4 marks]*
Level 3: An experiment to investigate force and acceleration which can be performed with the given equipment is fully described in detail. The answer is well structured. *[5 to 6 marks]*

Here are some points your answer may include:
Place all of the masses in the trolley.
Calculate and record the weight of the trolley.
Place the trolley on the starting line.
Release the trolley, so that it moves through the light gate, and record the acceleration measured.
Take one of the masses from the trolley, and attach it to the hook.
Calculate and record the new total weight of the hook and the new total weight of the trolley.
Reset the position of the trolley on the starting line.
Release the trolley again so that it moves through the light gate, and record the acceleration measured.
Repeat these steps until all the masses from the trolley have been moved to the hook.
Plot your results on a graph of acceleration against weight, and draw a line of best fit.

Page 120

Warm-Up Questions

1) The thinking distance is the distance travelled during your reaction time (the time between seeing a hazard, and applying the brakes).
2) The braking distance.
3) Any one from: e.g. poor grip on the roads increases braking distance / poor visibility delays when you see the hazard / distraction by the weather delays when you see the hazard.

4) Get the individual to sit with their arm resting on the edge of a table. Hold a ruler end-down so that the 0 cm mark hangs between their thumb and forefinger. Drop the ruler without warning. The individual must grab the ruler between their thumb and forefinger as quickly as possible. Measure the distance at which they have caught the ruler. Use $v^2 - u^2 = 2as$, $a = 9.8$ m/s^2 and $a = \Delta v \div t$ to calculate the time taken for the ruler to fall that distance. This is their reaction time.

5) Energy is transferred from the kinetic energy stores of the wheels to the thermal energy stores of the brakes.

Exam Question

1.1 stopping distance = braking distance + thinking distance,
So, braking distance = stopping distance
– thinking distance *[1 mark]*
At 40 mph,
stopping distance = 35 m (accept between 34 m and 36 m)
thinking distance = 13 m (accept between 12 m and 14 m)
[1 mark]
braking distance = 35 – 13 *[1 mark]*
$= $ **22 m** (Accept correct for above readings) *[1 mark]*

1.2 braking distance *[1 mark]*
The stopping distance is over twice as high as the thinking distance at 50 mph, so the braking distance must be bigger than the thinking distance.

1.3 Stopping distance is not directly proportional to speed *[1 mark]*. If stopping distance and speed were directly proportional, the relationship between them would be shown by a straight line / would be linear *[1 mark]*.

1.4 If the road were icy, the thinking distance graph would not change *[1 mark]* but the stopping distance graph would get steeper (as the braking distance would increase) *[1 mark]*.

The thinking distance graph doesn't change, because the icy road won't change your reaction time. But it will decrease the friction between the car and the road, so the braking distance increases.

Page 124

Warm-Up Questions

1) $p = mv = 2.5 \times 10 = $ **25 kg m/s**
2) In a closed system, the total momentum before an interaction must equal the total momentum after the interaction.
3) The momentum is zero.
4) It decreases the force.

Exam Questions

1.1 momentum = mass × velocity / $p = mv$ *[1 mark]*
1.2 $p = 60 \times 5.0$ *[1 mark]*
$= $ **300 kg m/s** *[1 mark]*
1.3 force = change in momentum ÷ time taken / $F = m\Delta v \div \Delta t$
Gymnast comes to a stop, so $m\Delta v = 300$ kg m/s
$F = 300 \div 1.2$ *[1 mark]*
$= $ **250 N** *[1 mark]*
2.1 momentum = mass × velocity / $p = mv$
$p = 650 \times 15.0$ *[1 mark]*
$= $ **9750 kg m/s** *[1 mark]*
2.2 momentum before = momentum after
Set direction of first car to be positive.
momentum of second car = 750 × (–10.0)
$= -7500$ kg m/s *[1 mark]*
Total momentum before = 9750 + (–7500)
$= 2250$ kg m/s *[1 mark]*
Total momentum after = (mass of car 1 + mass of car 2) × v
$2250 = (650 + 750) \times v$
so, $v = 2250 \div (650 + 750)$ *[1 mark]*
$= 2250 \div 1400$
$= 1.607142...$ *[1 mark]*
$= $ **1.61 m/s (to 3 s.f.)** *[1 mark]*

2.3 The crumple zone increases the time taken by the car to stop / change its velocity *[1 mark]*. The time over which momentum changes is inversely proportional to the force acting, so this reduces the force *[1 mark]*.

3 momentum before = momentum after
momentum of neutron before = 1 × 14 000
= 14 000
momentum of uranium before = 235 × 0
= 0 *[1 mark]*
momentum of neutron after = 1 × −13 000
= −13 000
momentum of uranium after = 235v *[1 mark]*
So, 14 000 = −13 000 + 235v *[1 mark]*
so, v = (14 000 + 13 000) ÷ 235
= 114.8936... *[1 mark]*
= **115 km/s (to 3 s.f.)** *[1 mark]*

Don't worry too much about the units in this question. The masses given are relative masses, with no units, so we couldn't use the standard units for momentum. As you're only looking for the velocity though, you can just do the calculation as normal, and make sure that the units on your final answer match the units for velocity given in the question.

Topic 6 — Waves

Pages 134-135
Warm-Up Questions

1) In a longitudinal wave, the vibrations are parallel to the direction of travel/energy transfer.
2) wave speed = frequency × wavelength / $v = f\lambda$
3) E.g. Set up and turn on a ripple tank. Then, dim the lights and turn on the lamp so a wave pattern can be seen on the screen below the ripple rank. From the screen, measure the distance between the shadow lines that are a certain number of wavelengths apart, e.g. ten wavelengths. Divide the distance by this number of wavelengths to find the average wavelength. Use the equation wave speed = frequency × wavelength / $v = f\lambda$ to find the speed of the waves.
4) true
5) E.g. Put one of the blocks on a piece of paper and trace round it. Shine a ray of light from a ray box onto one side of the block, at an angle to the block. Trace the incident ray and mark where it emerges from the block. Remove the block and draw a straight line (using a ruler) to connect the incident ray and the emerging ray. Draw the normal to the point of incidence and measure the angles of incidence and refraction. Repeat for the second block, making sure the angle of incidence is the same as for the first block. Compare the angles of refraction for the two blocks.

Exam Questions
1.1 transverse *[1 mark]*
1.2 5 cm *[1 mark]*
1.3 2 m *[1 mark]*
1.4 It will halve *[1 mark]*.
$v = f\lambda$, so if f doubles, then λ must halve, so that v stays the same.

2.1 specular *[1 mark]*
2.2

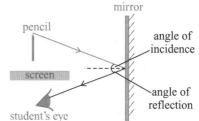

[1 mark for line representing reflected ray, 1 mark for the angle of incidence and the angle of reflection being equal, 1 mark for labelling the angle of incidence and the angle of reflection]

3.1 0.005 × 4 *[1 mark]* = 0.2 s *[1 mark]*
3.2 period *[1 mark]*
The period of a wave is the time taken for a full wave to pass a point.
3.3 One period is 8 divisions long,
so T = 0.005 × 8 *[1 mark]*
= 0.04 s *[1 mark]*
$f = \dfrac{1}{T} = \dfrac{1}{0.04}$ *[1 mark]*
= **25 Hz** *[1 mark]*

4.1 The distance he measures is 1 wavelength *[1 mark]*. This can be used, together with the frequency of the signal, in the formula for wave speed, wave speed = frequency × wavelength / $v = f\lambda$ *[1 mark]*.
4.2 wave speed = frequency × wavelength / $v = f\lambda$
So $v = 50 × 6.8$ *[1 mark]*
= **340** *[1 mark]* **m/s** *[1 mark]*

5.1 E.g.

[1 mark for refracting the ray towards the normal upon entering the prism, 1 mark for refracting the ray away from the normal as it leaves the prism and 1 mark for correctly labelling all the angles of incidence and refraction]

5.2 E.g. Place the prism on a piece of paper and shine a ray of light at the prism. Trace the incident and emergent rays and the boundaries of the prism on the piece of paper *[1 mark]*. Remove the prism and draw in the refracted ray through the prism by joining the ends of the other two rays with a straight line *[1 mark]*. Draw in the normals using a protractor and ruler *[1 mark]* and use the protractor to measure I and R at both boundaries *[1 mark]*.

Page 141
Warm-Up Questions
1) gamma rays
2) Alternating current is made up of oscillating charges, which produce oscillating electric and magnetic fields in the form of radio waves.
3) The microwaves penetrate a few centimetres into the food before being absorbed by water molecules. The energy from the absorbed microwaves causes the food to heat up.
4) E.g. to carry data over long distances.
5) It has enough energy to knock electrons off atoms — this can cause gene mutations, cell destruction and cancer.

Exam Questions

1.1 Any one of: e.g. flourescent light bulbs / tanning lamps

1.2 Any one of: e.g. sunburn / premature ageing of skin / blindness / increased risk of skin cancer *[1 mark]*.

1.3 A gamma ray emitter is injected into/swallowed by the patient *[1 mark]* and the gamma rays emitted are detected by an external detector *[1 mark]*. By seeing where the gamma rays come from, they can track the progress of the tracer around the body and check it is functioning correctly *[1 mark]*.

1.4 Gamma rays can pass out of the body without being absorbed *[1 mark]*.

2.1 Radio waves can bend around / pass through objects so the signal can reach the inside of the house *[1 mark]*. Light waves would be blocked by the mountain / walls, and so would not be able to reach the receiver inside the house *[1 mark]*.

2.2 Microwave radiation *[1 mark]*. It passes through Earth's watery atmosphere without being absorbed, so can reach the satellites *[1 mark]*.

2.3 They are transmitted through the atmosphere into space, where they are picked up by a satellite receiver orbiting Earth *[1 mark]*. The satellite transmits the signal back to Earth in a different direction, where it is received by a satellite dish connected to the house *[1 mark]*.

Page 147
Warm-Up Questions

1) Concave/diverging and convex/converging.

2) Concave lenses always create virtual images. Convex lenses can create real or virtual images.

3) It most strongly reflects the wavelengths of light corresponding to the red part of the visible spectrum (all other wavelengths are absorbed).

4) The wavelengths reflected from the object (those corresponding to the green part of the visible spectrum) are absorbed by the red filter. The lack of any visible light transmitted is perceived as black, so the object appears black.

Exam Questions

1.1 A concave lens causes parallel rays of light to diverge (spread out) rather than converge (come together) *[1 mark]*. This means that Ed's lens cannot focus the sunlight to start a fire *[1 mark]*.

1.2

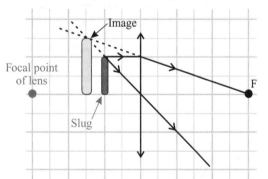

[1 mark for showing a ray going to a correctly positioned focal point, 1 mark for showing a ray going through the centre of the lens, 1 mark for showing both rays extended backwards (dotted) and image drawn where they cross]

You sometimes need to draw another focal point on the opposite side of lens, the same distance away from the centre line of the lens.

1.3 magnification = image height ÷ object height

= 1.5 ÷ 1 *[1 mark]*

= **1.5 *[1 mark — accept any answer between 1.4 and 1.6]***

With this question, it's easiest to use the grid squares as units and just count them. Alternatively, you could measure the lengths of the object and the image with a ruler.

2.1 Daylight consists of an even distribution of all the wavelengths *[1 mark]*. Light from a flat screen device consists of mostly wavelengths below 500 nm *[1 mark]*. It emits far fewer wavelengths greater than 500 nm *[1 mark]*.

2.2 E.g. use a colour (e.g. an orange) filter between the screen and the users eyes *[1 mark]* to absorb the blue light *[1 mark]*.

Pages 156-157
Warm-Up Questions

1) It emits more IR radiation than it absorbs.

2) Leslie cube

3) In a given place, more radiation is being emitted than is being absorbed.

4) A perfect black body is an object that absorbs all of the radiation that hits it. No radiation is reflected or transmitted.

5) longitudinal

6) Partial reflection is when some of the wave travels into the new medium (and is possibly refracted), whilst some is reflected.

7) E.g. earthquakes

8) P waves

9) They can't pass through the liquid outer core.

Exam Questions

1.1 Any one from: e.g. place the thermometers at equal distances away from the cube / place the thermometers at the same height as each other / make sure no thermometers are in direct sunlight/a draught *[1 mark]*.

1.2 C, A, B, D *[1 mark]*

Thermometer C will heat up fastest, since it is the one closest to the face which is most like a perfect black body and so the best infrared emitter.

1.3 The times recorded would be longer *[1 mark]*. Cooler objects emit infrared radiation at a lower rate *[1 mark]*.

1.4 E.g. the resolution of the digital thermometer is higher, so there will be less uncertainty in the results (the results will be more accurate) / the student is less likely to misread the temperature (human error is less likely) *[1 mark for each correct reason, up to a maximum of 2]*.

2.1 20 kHz / 20 000 Hz *[1 mark]*

2.2

[1 mark]

2.3 It will be partially reflected / some of it will be reflected and some will be transmitted into media 2 *[1 mark]*.

3.1 A *[1 mark]*. They only travel through solids *[1 mark]*, and so are stopped by the liquid outer core *[1 mark]*.

3.2 When the waves are following curved paths, it shows that the density of the material they are moving through is changing gradually *[1 mark]*, suggesting there are areas of the Earth's inner structure made up of a similar material of a changing density *[1 mark]*. When the waves abruptly change direction, it shows that the density at that point changes suddenly by a larger amount *[1 mark]*, suggesting they have entered a layer of a new material in the Earth's structure *[1 mark]*.

4.1 Bellatrix *[1 mark]*. It emits blue light which has a lower wavelength than red light *[1 mark]*. The peak wavelength of a hotter object is lower than the peak wavelength of a cooler object *[1 mark]*.

4.2 People on Earth wouldn't hear the explosion *[1 mark]* as sound can't travel through the vacuum of space *[1 mark]*.

Topic 7 — Magnetism and Electromagnetism

Page 162
Warm-Up Questions
1)

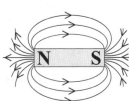

2)   It's made up of concentric circles around the wire (with the wire at the centre).
3)   E.g. cranes use them to pick up iron and steel in scrap yards.

Exam Questions
1.1   E.g. Put the magnets on a piece of paper and place many compasses in different places between the magnets to show the magnetic field at those points *[1 mark]*. The compass needles will line up with the magnetic field lines *[1 mark]*.
They could also use iron filings to show the pattern.
1.2   The field lines point straight across from the north pole towards the south pole *[1 mark]*.
1.3   Attraction *[1 mark]*, as opposite poles are facing each other and opposite poles attract *[1 mark]*.
2.1

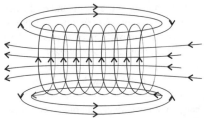

[1 mark for correct field shape, 1 mark for direction]
2.2   E.g. iron *[1 mark]*
2.3   The paperclips will fall *[1 mark]*, because an electromagnet only has a magnetic field if a current is flowing *[1 mark]*.

Page 166
Warm-Up Questions
1)   When a current-carrying wire in a magnetic field experiences a force.
2)   First finger — magnetic field
      Second finger — current
      Thumb — force (motion)
3)   Swapping the polarity of the dc supply (reversing the current), or swapping the magnetic poles over (reversing the field).

Exam Questions
1.1

[1 mark]
1.2   The direction of the force would be reversed too *[1 mark]*.
1.3   The force would be lower *[1 mark]*.
1.4   There would be no force on the wire *[1 mark]*.
2.1   E.g.   direction of rotation   axis of rotation

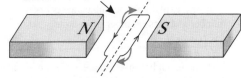

[1 mark for any indication that the current goes anticlockwise]

2.2   force = magnetic flux density × current × length of conductor inside field / $F = BIl$
      $F = 0.2 \times 15 \times 0.1$ *[1 mark]*
         $= \mathbf{0.3\ N}$ *[1 mark]*
2.3   By swapping the direction of the current/contacts every half turn (using a split-ring commutator) *[1 mark]* so the forces on the loop always act in a way that keeps the loop rotating *[1 mark]*.

Page 172
Warm-Up Questions
1)   Any two from: e.g. increase the speed of movement / rate of field lines cut / increase the strength of the magnetic field.
2)   Direct current.
3)   More turns on the secondary coil.

Exam Questions
1.1   Because electrical current is generated *[1 mark]*.
1.2   Any one from: e.g. move the magnet out of the coil / move the coil away from the magnet / insert the south pole of the magnet into the same end of the coil / insert the north pole of the magnet into the other end of the coil *[1 mark]*.
1.3   Any one from: e.g. push the magnet into the coil more quickly / use a stronger magnet *[1 mark]*.
1.4   Zero / no reading *[1 mark]*.
Movement is needed to generate a current.
2.1   A potential difference will not be induced in the secondary coil when using the dc supply *[1 mark]* because the magnetic field generated in the iron core is not changing *[1 mark]*. A pd is only induced if the secondary coil experiences a change in magnetic field *[1 mark]*.
2.2   power = potential difference × current / $P = VI$
         $= 12 \times 2.5$ *[1 mark]*
         $= \mathbf{30\ W}$ *[1 mark]*
2.3   Assuming that the transformer is 100% efficient, the output power is 30 W *[1 mark]*.
      power = potential difference × current / $P = VI$
      So, $I = P \div V$ *[1 mark]*
         $= 30 \div 4$ *[1 mark]*
         $= \mathbf{7.5\ A}$ *[1 mark]*
2.4 $V_s \div V_p = n_s \div n_p$
      So, $n_s = (V_s \div V_p) \times n_p$ *[1 mark]*
      $n_s = (4 \div 12) \times 15$ *[1 mark]*
         $= \mathbf{5\ turns}$ *[1 mark]*
3   When the ac current flows through the coil of wire in the magnetic field of the permanent magnet, the coil of wire experiences a force *[1 mark]*. The force causes the coil, and so the cone, to move *[1 mark]*. The alternating current is constantly changing direction and so the direction of the force on the coil is constantly changing and the cone vibrates back and forth *[1 mark]*. The vibrations cause pressure variations in the air that cause sound waves *[1 mark]*.

Topic 8 — Space Physics

Page 178
Warm-Up Questions
1)   Clouds of dust and gas/a nebula.
2)   A red super giant.
3)   E.g. planets and dwarf planets.
4)   gravity
5)   red-shift
6)   E.g. dark matter and dark energy.

Exam Questions

1.   Moons are **natural satellites** *[1 mark]*. They orbit **planets** *[1 mark]*. **Artificial satellites** *[1 mark]* are man-made, and most of them orbit **the Earth** *[1 mark]*.

2.1   Stars form from clouds of dust and gas which are pulled together by gravitational attraction *[1 mark]*. As the density increases, the temperature rises *[1 mark]*. When the temperature gets high enough, nuclear fusion happens and a huge amount of energy is emitted *[1 mark]*.

2.2   The forces acting on a main sequence star are balanced, so it doesn't collapse or explode *[1 mark]*. The energy released by nuclear fusion provides an outward force to balance the force of gravity pulling everything inwards *[1 mark]*.

2.3   They become unstable and eject their outer layer of dust and gases *[1 mark]* which leaves a hot, dense, solid core known as a white dwarf *[1 mark]*. White dwarfs then cool to become black dwarfs *[1 mark]*.

2.4   They start to glow brightly again, undergo more fusion, and expand and contract several times *[1 mark]*. Heavier elements are formed and the star eventually explodes in a supernova *[1 mark]*. The supernova leaves behind a neutron star or a black hole *[1 mark]*.

3   How to grade your answer:

Level 0:   There is no relevant information. *[No marks]*

Level 1:   There is a brief description of red-shift and the Big Bang theory. The answer lacks coherency. *[1 to 2 marks]*

Level 2:   There is a description of red-shift and the Big Bang theory, and some explanation of how red-shift provides evidence for the Big Bang theory. The answer has some structure. *[3 to 4 marks]*

Level 3:   There is a description of red-shift and the Big Bang theory, and a clear and detailed explanation of how red-shift provides evidence of the expansion of the universe, and so evidence for the Big Bang theory. The answer is well structured. *[5 to 6 marks]*

Here are some points your answer may include:

Red-shift is when the wavelengths of observed light from a source are longer than that of the light emitted by the source. This occurs when the source of light is moving away from the observer.

The light from most distant galaxies is observed to be red-shifted.

The light from galaxies that are further away from us is red-shifted more than the light from nearer galaxies.

This indicates that all of these galaxies are moving away from us and each other.

This suggests that the universe is expanding.

The Big Bang theory is the theory that the universe began as a small region of space that was very hot and dense, which exploded and has been expanding ever since.

Therefore, the observations of red-shift support the idea of the origin of the universe suggested by the Big Bang theory.

Practice Paper 1

1.1   liquid *[1 mark]*

1.2   The internal energy of the substance increases *[1 mark]* as energy is transferred by heating to the (kinetic and potential) energy stores of the particles *[1 mark]*.

1.3   melting *[1 mark]*

1.4   density = mass ÷ volume / $\rho = m \div V$ *[1 mark]*

1.5 $\rho = 0.36 \div (4.0 \times 10^{-4})$ *[1 mark]*
    = **900 kg/m³** *[1 mark]*

2.1   Strength: Any one from: e.g. the hot water in the beaker cools by conduction and convection, as in a hot water tank / the water can be raised to a similar temperature as in a hot water tank / both the tank and the model involve heat loss from water *[1 mark]*.
Weakness: Any one from: e.g. the mass of water used is much less than in a hot water tank / the insulation of a hot water tank wouldn't be cotton wool *[1 mark]*.

2.2   The thickness of the cotton wool jacket *[1 mark]*.

2.3   E.g. starting temperature of the water / length of time water is left to cool / volume of water used *[1 mark]*.

2.4   precise *[1 mark]*

2.5   The thicker the cotton wool jacket, the smaller the temperature change / the higher the final temperature of the water (and hence the less energy transferred from the water's thermal energy store) *[1 mark]*.

3   How to grade your answer:

Level 0:   There is no relevant information. *[No marks]*

Level 1:   There is a brief description of both models of the atom. The answer lacks coherency. *[1 to 2 marks]*

Level 2:   There is a description of both models of the atom, and some description of the scientific discoveries that led to the development of the nuclear model. The answer has some structure. *[3 to 4 marks]*

Level 3:   There is a clear and detailed description of both models of the atom, and of the scientific discoveries and experiments which led to the development of the nuclear model. The answer is well structured. *[5 to 6 marks]*

Here are some points your answer may include:

Model X is the plum pudding model of the atom.

The plum pudding model describes the atom as a sphere of positive charge, with negatively charged electrons within it.

Model Y is the nuclear model of the atom.

The nuclear model of the atom describes the atom as a nucleus, made up of positively charged protons and uncharged neutrons, orbited by electrons.

In the early 20ᵗʰ century (1909), Rutherford performed the alpha scattering experiment.

They fired a beam of positively charged alpha particles at a thin gold foil.

Based on the plum pudding model, they expected all the alpha particles to pass through the foil, with some deflection.

However, they found that most alpha particles passed through the foil without deflecting, while a small few were deflected back towards the emitter.

This suggested that most of the atom is empty space, since so many of the alpha particles passed through without deflecting.

It also suggested there was a small, positively charged 'nucleus' in the centre of the atom, which caused the backwards deflection of the alpha particles.

This formed the basis of the nuclear model of the atom.

Later, Niels Bohr adapted the nuclear model, when he discovered electrons orbited the nucleus at specific distances called energy levels.

In 1932, James Chadwick discovered the neutron, an uncharged particle with the same mass as a proton.

This led to the idea that the nucleus is made of protons and neutrons, explaining the imbalance between the atomic number and mass number of an atom.

4.1   gas *[1 mark]*

4.2   C *[1 mark]*

4.3   The line is flat because a change of state is occurring *[1 mark]* and all energy transferred to the substance is being used to break intermolecular bonds and change the state of the substance, and not increase its temperature *[1 mark]*.

4.4   thermal energy for a change of state = mass × specific latent heat / $E = mL$
So, $L = E \div m$ *[1 mark]*
$L = 69\ 000 \div 1.5$ *[1 mark]*
    = **46 000 J/kg** *[1 mark]*

5.1   Electrons have been transferred from the scientist's hair to the comb *[1 mark]*.

5.2   The hairs all have the same (positive) charge *[1 mark]*. Like charges repel, so the hairs repel each other and so stand on end *[1 mark]*.

5.3   (+) 0.2 nC *[1 mark]*
The hair will have an equal and opposite charge to the comb.

5.4   It will attract the hair *[1 mark]*.

6.1   The specific heat capacity of a substance is the energy required to raise the temperature of 1 kg of that substance by 1 °C *[1 mark]*.

6.2   change in thermal energy = mass × specific heat capacity × temperature change / $\Delta E = mc\Delta\theta$
So, $c = \Delta E \div m\Delta\theta$ *[1 mark]*
$\Delta\theta = 93 - 18 = 75$ *[1 mark]*
$c = 126\ 000 \div (1 \times 75)$ *[1 mark]*
    = **1680 J/kg°C** *[1 mark]*

6.3   Water has a higher specific heat capacity than oil *[1 mark]* so it transfers more energy to the surroundings compared to oil, for the same decrease in temperature *[1 mark]*.

7.1 $214 - 83 = $ **131 neutrons** *[1 mark]*

7.2
$^{222}_{86}$ Rn   *[1 mark]*

$^{210}_{83}$ Bi   *[1 mark]*

7.3   420 days *[1 mark]*

7.4   Half-life when activity drops to half original total.
Activity after 1 half-life = 80 ÷ 2 = 40 Bq
Using **Figure 9**, when activity = 40 Bq, time = 140 days.
Therefore, half-life = **140 days** *[1 mark]*

7.5   Alpha radiation is strongly ionising *[1 mark]*, and when it enters living cells it can easily kill them, or damage them and cause cancer *[1 mark]*. Alpha radiation is easily absorbed by thin barriers (e.g. skin) or the air, so is unlikely to reach the body's delicate organs if the source is outside the body *[1 mark]*. However it is much more dangerous inside the body, where it is almost certain to be absorbed by living cells and cause damage *[1 mark]*.

8.1   The current is higher in circuit A *[1 mark]*.
The total resistance is higher in circuit B *[1 mark]*.

8.2   power = potential difference × current / $P = VI$ *[1 mark]*

8.3 $P = 11 \times 0.30$ *[1 mark]*
    = **3.3 W** *[1 mark]*

8.4   As the intensity of light increases, the resistance of component C decreases *[1 mark]*.

8.5   E.g. a car engine temperature sensor *[1 mark]*.

8.6
Current in amps

Potential difference in volts

[1 mark]

9.1   Bio-fuels = 6×10^9 kWh *[1 mark]*
Hydro-electric = 1×10^9 kWh *[1 mark]*
[Allow 1 mark only if the answers are 6 kWh and 1 kWh and × 10⁹ is omitted]

9.2   Wind power *[1 mark]*

9.3   Any one from: e.g. there are more hours of daylight in summer than in winter, so more electricity can be generated from solar power / there are generally more clear days during summer so more electricity can be generated from solar power *[1 mark]*.

9.4   Advantage: Any one from: e.g. non-renewables are reliable / we can easily alter energy output to meet demand *[1 mark]*.
Disadvantage: Any one from: e.g. non-renewables will eventually run out / burning some non-renewables, e.g. fossil fuels, produces pollutants / burning some non-renewables, e.g. fossil fuels can cause global warming *[1 mark]*.

9.5   Any two from: e.g. they're not as reliable as non-renewables / they can have large initial setup costs / a number of power stations that use renewable resources can only be built in certain locations / a lot of renewable resources cannot easily have their power output increased to meet demand *[1 mark for each correct limitation]*.

9.6   Geothermal power requires a suitably volcanically active site to work, and there are few viable volcanic sites in the UK *[1 mark]*.

10.1

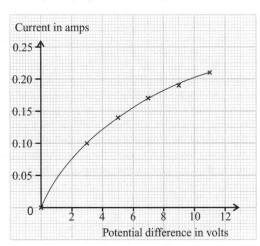

[2 marks for all points plotted correctly, otherwise 1 mark for any 3 plotted correctly]

10.2

Current in amps

[1 mark for curved line close to all points and through the origin]

10.3   As an electrical charge flows through the lamp, energy is transferred to its thermal energy store *[1 mark]*. As the current increases, more energy is transferred electrically to the thermal energy store of the component, increasing the temperature *[1 mark]*. Resistance increases with temperature *[1 mark]*, and so resistance increases with increasing current *[1 mark]*.

10.4 potential difference = current × resistance / $V = IR$
So, $I = V \div R$ *[1 mark]*
$= 15 \div 60$ *[1 mark]*
$= 0.25$ A *[1 mark]*
charge flow = current × time / $Q = It$
$Q = 0.25 \times 180$ *[1 mark]*
$= 45$ C *[1 mark]*

11.1 Earth wire: yellow and green (stripes) *[1 mark]*
Live wire: brown *[1 mark]*

11.2 The earth wire stops exposed metal parts / the casing of the device from becoming live in the event of a fault *[1 mark]*.

11.3 A live wire may still have a potential difference in it even when this circuit is broken / current is now flowing *[1 mark]*. If you touch the live wire, your body forms a link between the wire and the earth, and current flows through you, and can give you an electric shock *[1 mark]*.

11.4 power = (current)^2 × resistance / $P = I^2R$
$P = 10.0^2 \times 23.0$ *[1 mark]*
$= 2300$ W *[1 mark]*
energy transferred = power × time / $E = Pt$
$E = 2300 \times 180$ *[1 mark]*
$= 414\,000$ J *[1 mark]*

12 How to grade your answer:
Level 0: There is no relevant information. *[No marks]*
Level 1: There is a brief description of devices used in the national grid. The answer lacks coherency. *[1 to 2 marks]*
Level 2: There is a description of devices used in the national grid, and some explanation of how they are used to efficiently transmit electricity. The answer has some structure. *[3 to 4 marks]*
Level 3: There is a clear and detailed description of how devices such as transformers are used in the national grid to transmit electricity efficiently. There is also some mention of how the electricity is made suitable for home use after transmission. The answer is well structured. *[5 to 6 marks]*

Here are some points your answer may include:
The national grid needs to transmit high power electricity from power stations to consumers.
Because the power is high, and $P = IV$, the electricity must be transmitted with either high potential difference (V) or high current (I).
When a current flows through a conductor, work is done against its resistance and energy is dissipated to the thermal energy store of the conductor.
The higher the current, the more energy is transferred to thermal energy stores and wasted.
Transformers are a device which can change the potential difference of electricity without changing the power.
In the national grid, step-up transformers are used to increase the potential difference of the electricity before transmission.
Because the power is constant, increasing the potential difference of the electricity decreases the current.
So step-up transformers allow the electricity to be transmitted at very high potential difference and very low currents.
As it is transmitted at a very low current, very little energy is dissipated to thermal energy stores.
So transmission at low current is more efficient.
The transmission potential difference is too large to be used by consumers.
Before reaching the consumer, step-down transformers are used.
Step-down transformers decrease the potential difference back to a usable level for consumer appliances.

13.1 gravitational potential energy = mass × gravitational field strength (g) × height / $E_p = mgh$ *[1 mark]*
13.2 $E_p = 600 \times 9.8 \times 20$ *[1 mark]*
$= 117\,600$ J $= 120$ kJ (to 2 s.f.) *[1 mark]*
13.3 As the carriage drops, the energy is transferred from its gravitation potential energy store to its kinetic energy store, so the kinetic energy store graph slopes upwards as the g.p.e store graph slopes downwards *[1 mark]*. Due to conservation of energy, as there are no resistive forces, the sum of the energy in the gravitational potential and kinetic energy stores at any given point between A and B is constant (so the lines have the same, but opposite, gradient) *[1 mark]*.
13.4 kinetic energy = 0.5 × mass × (speed)^2 / $E_k = \frac{1}{2}mv^2$
Conservation of energy:
total energy before = total energy after
No resistive forces, so by conservation of energy, all energy transferred from g.p.e. store goes to the kinetic energy store.
So, $E_p = E_k = 117\,600$ J *[1 mark]*
So, $v = \sqrt{\frac{2E_k}{m}}$ *[1 mark]* $= \sqrt{\frac{2 \times 117\,600}{600}}$ *[1 mark]*
$= 19.79... = 20$ m/s (to 2 s.f.) *[1 mark]*

Practice Paper 2

1.1 X-rays are less easily absorbed by tissues, but are absorbed more by denser materials like bones and metal *[1 mark]*. The detector detects the X-rays transmitted at each point and creates a picture showing areas of tissue and bone *[1 mark]*.
1.2 X-rays are ionising radiation which can kill living cells or cause mutations and cancer *[1 mark]*.
1.3 Lead absorbs X-rays and prevents them reaching the radiographer *[1 mark]*.
1.4 Any two from: e.g. X-rays can be used to treat cancers / X-ray photographs can be used to diagnose bone fractures / X-ray photographs can be used to diagnose dental problems (problems with your teeth) *[1 mark for each correct answer]*.
2.1 E.g. frequency is the number of complete waves passing a certain point per second / the number of waves produced by a source each second *[1 mark]*.
2.2 wave speed = frequency × wavelength / $v = f\lambda$ *[1 mark]*
2.3 So, $\lambda = v \div f$ *[1 mark]*
$\lambda = 3.0 \times 10^8 \div 2.5 \times 10^{14}$ *[1 mark]*
$= 1.2 \times 10^{-6}$ m *[1 mark]*
2.4 Texture: Matt surfaces are better infrared emitters then shiny surfaces *[1 mark]*.
Colour: Black surfaces are better infrared emitters than white surfaces *[1 mark]*.
2.5 Any one from: e.g. to reduce the effect of random errors / to make the results more accurate *[1 mark]*.
3.1 The outward pressure/expansion due to nuclear fusion *[1 mark]* balances the inwards force due to gravitational attraction *[1 mark]*.
3.2 A white dwarf *[1 mark]*.
3.3 35 km/s *[1 mark]*
The smaller the orbit, the faster the orbital speed.
3.4 The object will weigh less on Venus than on Earth *[1 mark]* as the gravitational field strength of Venus is lower than Earth's (and weight is proportional to gravitational field strength) *[1 mark]*.
4.1 27 s *[1 mark]*
4.2 Graph is linear between 15 s and 27 s.
$27 - 15 = 12$ s *[1 mark]*
4.3 0 N *[1 mark]*
When an object is travelling at a constant speed in a fixed direction, all the forces are balanced and so the resultant force will be zero.
4.4 Between 9 m and 10 m *[1 mark]*.
The swimmer is travelling fastest when the gradient of the graph is steepest. Of the options given, the graph is steepest between 9 m and 10 m. You can draw a tangent to see this more clearly.

4.5   distance = speed × time / $s = vt$ *[1 mark]*

4.6 $v = s \div t$ *[1 mark]*
       = 20 ÷ 25 *[1 mark]*
       = **0.8 m/s** *[1 mark]*

4.7

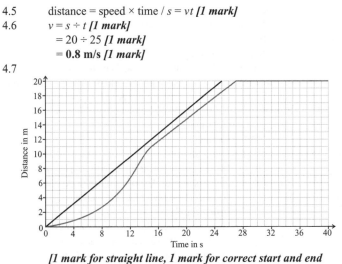

[1 mark for straight line, 1 mark for correct start and end point]

4.8   Yes, the camera will be able to film the swimmer for the whole length because its distance-time graph is always above the swimmer's *[1 mark]*, and so the camera is always ahead of the swimmer (or level at $t = 0$) *[1 mark]*.

5.1   Random errors *[1 mark]*

5.2*   How to grade your answer:

Level 0:   There is no relevant information. *[No marks]*

Level 1:   There is a brief description of an experiment using the equipment shown. The answer lacks coherency. *[1 to 2 marks]*

Level 2:   There is a description of an experiment which can be performed with the equipment shown, and some description of how the results should be processed to calculate the spring constant. The answer has some structure. *[3 to 4 marks]*

Level 3:   There is a clear and detailed description of an experiment which can be performed using the equipment show, and of how the to calculate the spring constant from the resulting force-extension graph. The answer is well structured. *[5 to 6 marks]*

Here are some points your answer may include:

Measure the mass of each of the masses using a mass balance.

Calculate the weight of each of the masses using $W = mg$.

Using the ruler, measure the length of the spring when it has no masses hanging from it (the unstretched length).

Hang a mass from the spring, and record the force applied by the mass (the weight of the mass) and the new length of the spring.

Calculate the extension of the spring by subtracting the unstretched length from the new length.

Repeat these steps, increasing the mass hanging from the spring, recording the new weight and calculating the extension each time.

After you have a suitable number of points, plot your results on a force-extension graph, with force on the y-axis, and tension on the x-axis.

Draw a line of best fit on your results.

Identify the linear part of your graph.

Since $F = ke$, $k = F \div e$, the gradient of the linear part of the graph is equal to the spring constant.

Calculate the spring constant by calculating the gradient of the linear part of the graph.

5.3   The limit of proportionality *[1 mark]*.

5.4   Below the limit of proportionality (where the graph is linear) the spring obeys $F = ke$.
So, k = gradient = change in y ÷ change in x
change in y = 7.0 – 0 = 7.0 N
change in x = 3.5 – 0 = 3.5 cm = 0.035 m
k = 7.0 ÷ 0.035 = **200 N/m**
[3 marks, otherwise 1 mark for an attempt to calculate the gradient of the linear part of the graph and 1 mark for the correct units]
You'd get the marks here for using any values for the change in y and x, as long as they're from the linear part of the graph.

6.1   The alternating current in the primary coil causes a changing magnetic field in the iron core and so in the secondary coil *[1 mark]*. The changing magnetic field induces an alternating potential difference in the secondary coil (the generator effect) *[1 mark]*.

6.2   E.g.

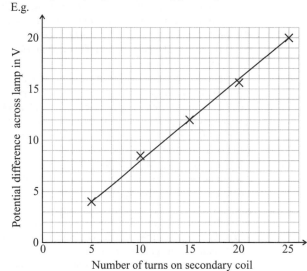

[1 mark for appropriate line of best fit]

6.3   A step-up transformer *[1 mark]*

6.4 $V_p \div V_s = n_p \div n_s$
So, $n_s = (V_p \div V_s) \times n_p$ *[1 mark]*
$n_s = (400\,000 \div 25\,000) \times 5000$ *[1 mark]*
     = **80 000** *[1 mark]*

6.5 $V_s I_s = V_p I_p$
So, $I_p = (V_s \times I_s) \div V_p$ *[1 mark]*
$I_p = (400\,000 \times 250) \div 25\,000$ *[1 mark]*
     = **4000 A** *[1 mark]*

7.1   50° (Allow values between 48° and 52°) *[1 mark]*

7.2

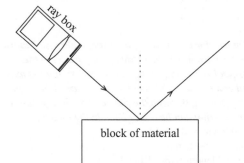

block of material

[1 mark for straight line drawn from point of incidence, 1 mark for angle with the normal equal to the angle stated in 7.1]

7.3   The speed of light is slower in the block than in air *[1 mark]*.

8.1   E.g. Reaction times vary from person to person *[1 mark]*. A sample needs to be large to be representative of the population *[1 mark]*.

To get a single representative value, it's a good idea to take an average from a large number of results.

8.2   Any one from: e.g. the condition of the tyres / the weather / the condition of the road surface / the condition of the brakes *[1 mark]*.

8.3 $v^2 - u^2 = 2as$
So, $a = (v^2 - u^2) \div 2s$ *[1 mark]*
   $= (0 - 30^2) \div (2 \times 100)$ *[1 mark]*
   $= -4.5$ m/s^2
So, deceleration = **4.5 m/s^2** *[1 mark]*

Since the question asked for deceleration, you should ignore the minus sign when you quote your answer.

8.4   The steering column collapsing slows the driver down more gradually. This increases the time over which the change of momentum happens *[1 mark]* which reduces the forces acting on the driver *[1 mark]* since force is equal to the rate of change of momentum *[1 mark]*.

9.1   The motor effect *[1 mark]*

9.2   The motor effect causes a current-carrying wire in a magnetic field to experience a force, so when current flows in the coil, the wire on each side feels a force *[1 mark]*. The force on one side of the coil acts upwards, and the force on the other side acts downwards, so the coil rotates, and turns the attached fan blades *[1 mark]*.

9.3   The metal bar won't feel any force *[1 mark]* because the current through it is parallel to the magnetic field *[1 mark]*.

9.4   Sound waves hit the diaphragm that is attached to a coil of wire *[1 mark]*. This causes the coil of wire to move back and forward in the magnetic field *[1 mark]*, which generates an alternating current in the coil *[1 mark]*. The generated current depends on the properties of the sound wave, so variations in sound are converted into variations in current *[1 mark]*.

10.1   Convex lens *[1 mark]*

10.2

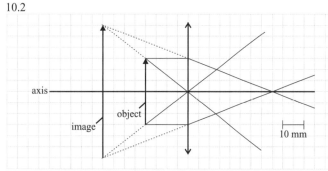

[3 marks for complete ray diagram, otherwise 1 mark for a ray from the object travelling parallel to the axis and refracting so that it travels through the axis and can be traced back to the image, and 1 mark for a ray passing through the centre of the lens without bending that can be traced back to the image.]

10.3   focal length = 8 squares,
2 squares = 10 mm
so focal length = $(8 \times 10) \div 2$
          = **40 mm** *[1 mark]*

10.4   object height = 6 squares = 30 mm
image height = 12 squares = 60 mm
magnification = image height ÷ object height
          = $60 \div 30$ *[1 mark]*
          = **2** *[1 mark]*

You could also have used the heights in squares instead. As long as the image and object heights are in the same units, you'll get the same answer.

11.1   moment = force × (perpendicular) distance from the pivot / $M = Fd$ *[1 mark]*

11.2 $M = 100 \times 0.2$ *[1 mark]*
      = **20** *[1 mark]* **Nm** *[1 mark]*

11.3   The moment will decrease *[1 mark]* because the perpendicular distance between the line of action of the force and the axis of rotation/pivot/point A will decrease *[1 mark]*.

12.1   weight = mass × gravitational field strength / $W = mg$ *[1 mark]*

12.2 $W = 80.0 \times 9.8$ *[1 mark]*
      = **784 N** *[1 mark]*

12.3*   How to grade your answer:
Level 0:   There is no relevant information. *[No marks]*
Level 1:   There is a brief description of the skydiver's motion. The answer lacks coherency. *[1 to 2 marks]*
Level 2:   There is a description of the skydiver's motion, with brief reference to the forces acting on him. The answer has some structure. *[3 to 4 marks]*
Level 3:   There is a clear and detailed description of the skydiver's motion, including details of the forces acting on him. The answer is well structured. *[5 to 6 marks]*

Here are some points your answer may include:
At first, the gradient of the graph is steep, showing that the skydiver is accelerating quickly.
This is because the force of gravity acting on the skydiver is much more than the air resistance slowing him down.
The graph then starts to level off, showing that acceleration is decreasing.
This is because the air resistance increases with speed, so the resultant force acting on the skydiver decreases.
Eventually, the air resistance becomes equal to the accelerating force of weight.
At this point, the skydiver travels at a constant (terminal) velocity.
This is represented by the straight horizontal line on the graph.

12.4   Area under the graph = ~16 squares
(Allow between 15 and 17) *[1 mark]*
Each square = $30 \times 10 = 300$ m
Distance fallen = 16×300 *[1 mark]*
          = **4800 m**
          [1 mark, allow between 4500 and 5100]

Index

A

absorption 130, 148, 151
  of infrared 148
  of light 130, 145, 146
acceleration 105, 111
  due to gravity 18, 109,
      113, 117, 176
  measuring 113, 182
  on distance-time
      graphs 106
  on velocity-time
      graphs 107
  uniform 105
accuracy 6
activity (radioactivity)
      77, 78
air bags 123
air resistance 108, 109
alpha radiation 75, 76
alpha scattering
      experiment 72
alternating currents (ac)
      52, 136, 169, 171
alternators 169
ammeters 42, 183
amplitude 126
angle of incidence
      130, 131
angle of reflection
      130, 133
angle of refraction
      130, 131
anomalous results 7
area under a graph 96, 107
artificial satellites 175
atmosphere 102, 151
atmospheric pressure
      68, 102
atomic models
  development of 72, 73
  nuclear 73
  plum pudding 72
atomic number 74, 76
atoms 72-76
averages 8

B

background radiation 81
bar charts 9
beta radiation 75, 76
bias 2
Big Bang theory 177
bio-fuels 35
black body radiation 150
black dwarfs 174
black holes 174
boiling 65, 66
brakes 116, 118
braking distances 116, 118

C

cancer 83, 139, 140
carbon neutral 35
categoric data 9
cavity wall insulation 26
centre of mass 88
chain reactions 84
changes in momentum 123
changes of state 65, 66
charge
  electric 40, 54, 60
  ions 73
  of a nucleus 72-74, 76
  relative charges of
      particles 73
  static 58, 59
charged spheres 60
circuit diagrams 40
circuit symbols 40
circuits 40, 42, 43, 46-50
circular motion 176
closed systems 17
coal 31, 36
colour filters 146
colours 145, 146
compasses 159, 160
components (electrical)
      40, 43
components (of a force) 91
compression (of springs)
      93, 94
compressions (in waves)
      127, 152
concave lenses 142, 144
conclusions 13
condensing 65, 66
conduction 24
conservation of energy 23
conservation of mass 65

conservation of momentum
      121, 122
contact forces 87
contamination 82
continuous data 9
control rods 84
control variables 5
convection 26
convection currents 26
converting units 12
convex lenses 142-144
correlations 10, 14
cosmic rays 81
count-rate 77
crash mats 123
crumple zones 123
current 40-43, 48, 49
  alternating
      52, 136, 169, 171
  direct 52, 169
  in parallel 49
  in series 48
  I-V characteristics 43
  magnetism 160,
      163-165, 167-169
  measuring 183
current-carrying wires
      160, 163, 165

D

dangers of ionising radiation
      82, 83, 139, 140
dark energy 177
dark matter 177
deceleration 105, 118
  on distance-time graphs
      106
  on velocity-time graphs
      107
density 64, 100, 101
dependent variables 5
designing investigations 5-7
diffuse reflection 130, 133
diodes 41, 43
direct currents (dc) 52, 169
discrete data 9
displacement (distance) 104
displacement (waves) 126
dissipated energy 23, 26
distance 104
distance-time graphs 106
double-glazing 26
drag 108, 109
draught excluders 26
dwarf planets 175
dynamos 169

E

ear drums 152
earth wires 52
earthquakes 155
Earth's structure 155
Earth's temperature 151
echo sounding 154
efficiency 28
elastic deformation 93, 96
elastic objects 93
elastic potential energy
      stores 19, 93, 96
electric cars 37
electric field strength 60
electric fields 60
electric heaters 28, 138
electric motors 165
electric shocks 52
electricity 40-43, 45-50,
      52-56, 58-60
  generation 167-169
  supply and demand 55
  usage 36
electromagnetic induction
      167-171
electromagnetic spectrum
      136
electromagnetic waves 136
  black body radiation 150
  dangers 140
  gamma rays 75, 76,
      139, 140
  infrared 138, 148
  microwaves 137
  uses 136-139
  UV 139, 140
  visible light 132, 139,
      143-146
  X-rays 139, 140
electromagnetism 161
electrons 58, 72, 73, 75
electrostatic attraction
      59, 60
electrostatic repulsion
      59, 60
elements 74, 76, 174
emission (infrared)
      148, 149
energy 17-21, 23-28,
      31-37
  conservation 23
  internal 65, 68
  stores 17, 19
  transfers 17, 18, 24-27

Index

Index

Index